East Texas Calling:

The Centennial History of the Lufkin-Conroe Telephone Exchange

By BOB BOWMAN

ISBN 1-878096-49-4

First Printing, 1998

Best of East Texas Publishers
PO 1647
Lufkin, TX 75902
Phone (409) 634-7444

Dedication

To the employees
of the Lufkin-Conroe Telephone Exchange, Inc.
for their unselfish dedication and hard work during 100 years.

Table of Contents

Introduction

In 1898, Dr. Alexander Madison Denman and attorney Edwin James Mantooth strung a wire between their offices in downtown Lufkin and installed Lufkin's first telephones. Two years later, they founded a telephone exchange to extend service to others in the community.

In 1998, the Lufkin-Conroe Telephone Exchange, Inc., the result of their pioneer telephone system, celebrates its 100th year in East Texas.

With its services focused principally in the Lufkin and Conroe areas—two of the fastest growing communities in Texas—the Lufkin-Conroe Telephone Exchange is not only the product of the Denman and Mantooth families' commitment to East Texas, but a tribute to the hundreds of men and women who have worked for the company since its inception, shaping Texas' fourth largest telephone company—and the nation's 20th biggest non-Bell system.

This book, which tells the story of 100 years of telephone service in East Texas, is an effort to capture for posterity the noble effort that went into the creation of the company, the dedication of four generations of Denmans and Mantooths, and the contributions of the company's employees in Lufkin, Conroe, Alto, Montgomery, Wells and other communities where LCTX provides services.

The company appreciates the work of Bob Bowman of Lufkin for his research and writing of "East Texas Calling: The Centennial History of the Lufkin-Conroe Telephone Exchange.Inc.," We are also grateful to Maggie Jessup of Conroe for her additional research.

Ironically, as the last pages of "East Texas Calling" were being written, the company embarked on a new chapter in its history. In August of 1997, the Board of Directors, comprised largely of fourth generation members of the Denman and Mantooth families, agreed to merge the company with Texas Utilities, Inc., one of the nation's most respected companies. This exciting move will thrust the company into even newer telecommunications directions in the years ahead.

As you read "East Texas Calling," we hope you will learn some valuable lessons:

- The value of the American enterprise system.
- The worth of hard work.
- The need to recognize people for their contributions.
- And the importance of being a part of your hometown.

We hope you enjoy the history of our company. All of us associated with the Lufkin-Conroe Telephone Exchange, Inc. had a wonderful time making it.

G.I. (Larry) Ross
Chairman of the Board, President and Chief Executive Officer
Lufkin-Conroe Communications, Co.

Virgin forests such as this were found in Angelina County in the 1880's when Lufkin was founded.

1 An idea comes to Lufkin

The clatter of the railroad awakened Dr. Alexander Madison Denman from his nap in the jostling coach car.

It had been a long journey, first from Lufkin, Texas, northward to Shreveport, Louisiana over the Houston, East and West Texas Railroad and then north over a connecting railway for Chicago and the 1893 World's Columbian Exhibition, a world's fair coinciding with the 400th anniversary of the discovery of America.

Now, on the return trip to Lufkin, Denman had something on his mind. At Chicago, he had witnessed a demonstration of Alexander Graham Bell's new invention, an apparatus he called "the telephone."

Dr. A. M. Denman

Denman was intrigued by inventors and new technologies. In 1882, he and his father, Colonel Lafayette Denman, had been instrumental in convincing railroader Paul Bremond to run his HE&WT line through Denman Springs, an indistinct settlement of a few families in central Angelina County.

Denman was also on the verge of building Lufkin's first hospital. It had been on his mind for some time and he had already begun sketching plans for a three-story building that would serve as his family's home, a medical office, and a hospital.

At the Chicago fair, Denman visited a display by the Bell Telephone Company, which had been founded by the Scottish inventor in 1877. Bell had studied the possibilities of electrical communications as an 18-year-old in Scotland, moved to Canada in 1870, and found work in the U.S., training teachers in the use of his father's system of visible speech and tutoring deaf and deaf-mute pupils.

In 1876, while testing an electrical communications apparatus in his laboratory, Bell upset a jar of battery acid on his work bench, and called his assistant, Thomas A. Watson, "Mr. Watson, come here, I want you." The apparatus on the bench carried his words to Watson in another room, and the telephone was born.

Now, in 1893, Denman had "the telephone" on his mind. He saw Bell's communications device as a valuable tool for saving time, reducing much of the necessity for horse-and-buggy travel, and linking his medical practice with his patients and friends. He was intrigued with the idea of bringing the telephone to

Lufkin.

When the HE&WT jerked to a halt beside Lufkin's depot on Cotton Square, Denman gathered his luggage and walked several blocks to his medical office. He greeted his family and checked on his patients, and then sat down to contemplate the telephone.

One of his first visits in the next few days was to his old friend, Judge Edwin James Mantooth, an attorney. They had shared the experiences of building Lufkin, investing in many of its early businesses, and organizing the civic improvements that would, in turn, shape Lufkin's evolution.

Judge E. J. Mantooth

Like Denman, Mantooth was a visionary. He helped found one of Lufkin's first business firms, Mantooth and Bonner, served as an alderman and city secretary when the city was incorporated in 1890, and had been one of the forces behind moving the Angelina County courthouse to Lufkin in 1892.

Mantooth found Denman's enthusiasm for the telephone contagious. The concept was simple, but promising. Together, they agreed to learn more about Bell's invention with the goal of bringing it to Lufkin. In 1898, convinced the technology was promising, they ordered two telephones, which arrived by railroad a few months later. The telephones were soon installed in Denman's medical office and Mantooth's law office.

An early telephone of the type used around the 1900's

The equipment they bought was probably a simple system sometimes called "a series telephone," which was designed for use on lines where only two instruments were installed. The price of the system, according to an early Sears, Roebuck and Company catalog, was $7.80 for each telephone. Each telephone weighed 41 pounds and came with two dry batteries.

In all likelihood, Denman and Mantooth did not use an actual Bell telephone. Following the expiration of the patents on the original Bell telephone, the 1880s were marked by the emergence of a large number of telephone companies and equipment manufacturers.

Using a telephone of the type installed by Denman and Mantooth was easy. A user turned a magnetic crank at one end of the line, creating an electrical current that caused another phone at the other end of the line to ring. To answer, the person at the other end picked up the receiver and said, “Hello.”

Within days of the telephone’s first calls, Lufkinites spread the word of Denman and Mantooth’s acquisition. Their friends dropped into their offices to see the telephone at work and for days, the two men spent most of their spare time talking to each other to demonstrate the instrument’s capabilities to other businessmen and professionals.

Lufkin, everyone concluded, was on the edge of an exciting new technology. The railroad community of a thousand people would never be the same.

Dr. A. M. Denman's home at the corner of Howe and Angelina Streets in Lufkin.

2 The Denmans

If any single family deserves the reputation as Lufkin's founding family, it is the Denmans.

In the early 1880s, when Paul Bremond, president of the Houston, East and West Texas Railroad, decided to lay his narrow-guage rails through Angelina County, he set into motion a series of events leading to Lufkin's creation.

The county seat at the time was Homer, where the majority of the county's lawyers, physicians and other professional people lived. The county's first established church, the First Methodist Church, was built there and the county's principal businesses were clustered around the courthouse square. Dr. W.W. Manning—a physician, druggist and sawmill operator who had in 1858 been responsible for moving the courthouse from Jonesville—headed a group of influential men who made most of the political decisions affecting Angelina County.

Before he began building Homer, Dr. Manning had lived at Marion, the first county seat, and had served as Marion's postmaster.

Historian W.J. Townsend described Dr. Manning as "an intelligent, progressive and useful citizen...who did much to promote its general welfare." Manning founded the county's first mechanical sawmill, owned the first piano, and imported the first jersey milk cattle and pure-bred chickens.

At Homer, Dr. Manning surrounded himself with men of like caliber. Among them was Colonel Washington Lafayette Denman, who served as Homer's second postmaster between 1858 and 1866.

Encouraged by men like Denman, Dr. Manning worked hard to make Homer—the namesake of his hometown in Louisiana—an island of culture in frontier Angelina County, which had a reputation for shootouts and rowdiness. A number of men were shot down on Homer's streets during the town's zenith and at one time four saloons operated near the courthouse. One of the gunshot victims was Sheriff W.R. McMullen, who was killed by a stray bullet during a family shootout.

It is ironic, therefore, that Homer—which sought to become the fountainhead of culture and business in Angelina County —should be best remembered for a rowdy, drunken brawl that supposedly killed the town and founded its archrival.

According to a story handed down by generations of Lufkinites, including the Denmans, Lufkin was little more than a dirt road in a forest clearing in 1881 when a surveying crew from Bremond's Houston, East and West Texas Railroad came into Angelina County to plot the railroad's line through East Texas. The crew supposedly began its work by surveying a route that would take the railroad through Homer, the county seat.

The story goes that the crew decided to spend Saturday night at one of Homer's saloons. As the night wore on, their patronage became too rowdy for Constable W.B. (Buck) Green, who put the crew into jail. The next morning, the men paid their fines and were freed.

The Dr. A. M. Denman family from left to right: Byford Denman, Nell Denman Abram, Linwood Denman, Kester W. Denman, Lillian Denman Shands, Archie Denman Hucherson, Peyton R. Denman and Mrs. Biddie Walker Denman. Dr. Denman died before this photo was made.

Green's actions, however, infuriated the chief of the survey crew, a man that some accounts identify as Edwin P. Lufkin. He supposedly ordered the crew to retrace its steps and plot a new route for the railroad, one that would bypass Homer.

The new route chosen conveniently ran through land owned by Colonel Denman and his son, Dr. Alexander Madison Denman, who reportedly had hosted the survey crew at their home a few days earlier.

When rumors of a new town on the railroad began to be whispered in Angelina County, Colonel Denman, Dr. Denman, and P.C. Abney evidenced their willingness to donate land for the townsite and a railroad depot. Jim McMullen, writing in the January 27, 1935, issue of *The Lufkin Daily News*, wrote: "The railroad engineers were staying at the home of P.C. Abney, there being no boarding house in the vicinity, and no doubt they were enjoying numerous social occasions as there were three young ladies, daughters of Mr. Abney, in the home...how much the charming surroundings had to do with the engineers' decision is not known, but soon a townsite was surveyed and business houses began to be built."

At the time, according to McMullen, Colonel Denman, Abney and Dick Abney were the only residents of the community. The railroad depot was located on property donated by Colonel Denman and the town began to grow around the railroad line. What is now Abney Street was the town's only street.

In 1860, Colonel Denman bought a third acre of land, Lot 1, Block 12, of the town of Homer, but moved to Denman Springs on the old Davisville Road, about three and a half miles from the present-day site of Lufkin. Deed records indicate Denman purchased one piece of land after another until he owned most of the land from Denman Springs, along Peach Creek, through the Moffett community to the Angelina River. He also owned other large parcels of land in what is now the southern part of Lufkin. In 1860, the U.S. census said he owned $5,000 in real property and $18,270 in personal property.

Later, Denman acquired a home on North First Street in Lufkin, where he and his wife reared and educated their children.

In 1890, when Lufkin's leadership decided to incorporate the eight-year-old settlement, the city's boundaries were described as "...beginning at the head of Brushy Creek near H. Rhodes' place to include same, then to where C. Mantooth lives to include him, thence to the southwest corner of W.C. Denman's homestead land, thence to the southeast corner of Charlie Harrell's homestead land, then to Denman Springs on Brushy Creek, thence up the creek to the beginning..."

Dr. A. M. Denman

Colonel Denman was born November 23, 1830, the son of Felix Gilbert Denman and Nancy Hutchison. His birthplace was probably in Franklin County, Georgia, although some relatives believe he was born in Cass County.

Little is known of Colonel Denman's childhood, but he probably worked on his father's farm. His later life reflects an understanding of farming and he appeared to be well-educated.

As a young man, he apparently succumbed to the lure of Texas and joined a group of Georgians who came to Texas to seek land and opportunity. His descendants say that Denman came to Texas with his uncle, William Storm Denman, in 1843, seven years after Texas won its independence from Mexico at the Battle of San Jacinto. He apparently returned to Georgia in 1849, where he married Algeline Swaggerty of Pigeon River, Cock County, Tennessee, and came back to Texas in the early 1850s, finally settling near Homer in Angelina County in 1853. In 1858, he was appointed by the Texas Legislature to serve on a Board of Examiners responsible for examining school teachers in Angelina County. That same year he was also elected a county commissioner.

When he was 31, while still a county commissioner, Denman volunteered for the Confederate Army in the Civil War. Although Angelina County voted against secession, a company of men was organized in the county to fight for the

Mrs. A. M. Denman in later years.

South. Company D of the Seventh Texas Cavalry was mustered at San Antonio and was designated as the Third Regiment of General Sibley's brigade. While Denman's name did not appear on the muster roster, enlistment papers from Washington, D.C. show that he was a first lieutenant in the Texas Cavalry, Anderson's Brigade, Company E.

Denman was placed in command of his regiment in Louisiana, where he served with distinction and was rewarded with the rank of colonel. He was given a medical discharge at Shreveport, Louisiana, on August 29, 1864, with an eye disease, which caused him frequent attacks of neuralgia.

After the Civil War, Denman returned to Homer, where he was known as a dynamic, forceful and persistent individual. He was a successful lawyer and served in a number of political offices.

Colonel Denman was a staunch Democrat. He had a dislike for carpetbaggers, Yankees, and Republicans. He was elected to the 14th Texas Legislature to represent Angelina and Nacogdoches counties after a hard-fought battle that resulted in a shooting incident at Homer, the county seat. The campaign had resulted in hard feelings between the Denmans and the Bonner family and, on the day of the shooting, Charles Bonner told his family before leaving home, "One of us will eat breakfast in Hell in the morning."

Arriving at Homer, Bonner and B. Carter, who was engaged to Colonel Denman's daughter, Sarah Ann, killed each other, and Bonner's father was wounded in the groin.

Colonel Denman was elected to a second term in the Texas Legislature by a 2,000-vote majority and completed his political service in 1882. The weekly *Austin Democratic Statesman* of June 3, 1876, said of Denman: "Mr. Denman served with much zeal and considerable ability. He was a fearless, sometimes reckless debater, and occasionally got off witticisms highly appreciated by his audience. Mr. Denman is a gentleman of kind feelings and warm friendships and possesses often a degree of tolerance that is commendable."

Colonel Denman was also a rugged individualist who enjoyed liquor, but was temperate and seldom given to drunkenness. On the 14th day of the 17th Texas Legislature in 1881-82, he rose to a point of order and accused *The Houston Post* of having wrongly reported him as being drunk, demanding a retraction from the newspaper.

Colonel Denman and his wife Algeline, known affectionately as Algie, had seven children. They included:

• Sarah Ann Texas Denman, who married Dr. Barclift and later Edward Clay Shirey of Georgia. Family legend says Sarah Ann was given the name Texas by her parents while they were enroute to Texas. They had hoped Sarah Ann would be born in Texas, but the birth occurred just before they crossed into Texas.

• W. Clark Denman, who married Sue Davis. He was the owner of Lufkin Livery Stable.

• Mariette (Molly) Denman, who married John Byrnes.

• Lydia (Lyde) Bragg Denman, who married Mac McCarty and organized the Baptist church near the Denman home at Denman Springs. After Mr. McCarty died, Lyde married a Mr. Hubbard.

- Richard (Dick) Lafayette Denman, who married Anna Katherine Jones.
- Felix Grundy Denman, who lived only six months.
- Alexander Madison Denman, who married Mary Caroline (Biddy) Walker.

Algeline Denman died November 1, 1890, and was buried in Walker Cemetery. In his loneliness, Colonel Denman remarried, but the second marriage lasted only a few months and ended in divorce, followed by his death on December 18, 1891.

Alexander Madison Denman, often called "Dr. Mat," was a graduate of Roanoke College in North Carolina. He became a well-known doctor and surgeon in Angelina County and built Lufkin's first hospital, a three-story structure that served as his home and medical office, behind his father's home on Angelina Street in Lufkin. The hospital stood a few blocks north of the HE&WT railroad and faced the railroad tracks.

Dr. Denman was a progressive, civic-minded physician who embraced new technology, as evidenced by his interest in the telephone in 1898 and his installation of Lufkin's first X-ray machine, as well as the town's first surgery suite, in his pioneer hospital in 1905. He was Lufkin's mayor in 1903-1905.

Dr. Denman was registered as a physician in Angelina County as early as 1886—only four years after Lufkin was founded—and in 1890 he was one of seven practicing physicians. The others were R.L. Denman, a brother; J.A. Abney, A. Brown, H.B. Chancey, A.H. DuBose and R.B. Warren. Seven physicians, including Dr. A.M. and R.L. Denman, were also listed in 1893, but some of the names had changed. The other five were A. Brown, Amasa H. DuBose, T.W. Largent, Lafayette Mantooth, and W.N.M. Smith.

Both A.M. and R.L. Denman graduated from the University of Tennessee and began their practice in Lufkin in 1886 and 1889.

Born July 31, 1858, Dr. A.M. Denman died October 1, 1908, a week after his automobile, the first on Lufkin's streets, was struck late at night by a flatbed railroad car which had been bumped from a train during a coupling some distance from the crossing. There were no railroad crossing barriers or warning devices at the time and the railcar could not be seen or heard. Dr. Denman and his nurse were enroute to treat a patient in Lufkin. The nurse was uninjured, and Dr. Denman died from an infection secondary to a bladder injury sustained in the accident. It was not only Lufkin's first automobile accident, but the first involving a train and a motorized vehicle.

Dr. Denman and his wife, Mary Caroline (Biddy) Walker, raised seven children, plus nephews and nieces. Their children were:

- Peyton R.
- Archie Lovell, who married Dave C. Hutcherson.
- Olive Lillian, who married Ned D. Shands.
- Kester Walker, who married Cleo Lydia Mantooth.
- Linwood Huchinson, who married Ruth Clovis Richardson.
- Mary Nell, who married Henry H. Abram.
- Byford Harvey.

Three of the children—Peyton, Linwood, and Byford—followed in their father's footsteps and became physicians and another child, Kester Walker Denman , became an attorney with his father-in-law, Judge E.J. Mantooth. The law firm became the seat of power for the Denman and Mantooth families' telephone interests in the formative years.

The Mantooth home in Lufkin following a snow storm. Left to right are Kester W. Denman, Jr., Caledonia Mantooth and E. J. Mantooth.

3 The Mantooths

When Dr. Alexander Madison Denman brought the idea of the telephone to Lufkin, the man with whom he shared his enthusiasm was an old family friend—and a visionary like himself.

Judge Edwin James Mantooth, like Dr. Denman, already had a reputation for progressiveness, foresight, and business acumen.

Born April 10, 1852, in Cock County, Tennessee, Judge Mantooth was the son of Thomas J. Mantooth, who came to Angelina County from Tennessee by wagon in 1857, settling on the Neches River. He was a cattleman, farmer and chief justice (county judge) of Angelina County from 1858 to 1862.

Mantooth was elected chief justice in 1858, succeeding Joseph Herrington. Mantooth was in office when a large number of the county's residents petititoned the county to remove the county seat of justice from Jonesville to Homer. Mantooth set a November 22, 1858, election, and a 147-vote majority sent the courthouse to Homer. Mantooth remained in the chief justice office until 1862 and also served as probate judge. He opposed secession from the Union until Texas joined the Confederacy and then lent his moral support to the South.

Thomas J. Mantooth's children, by his first wife Mary Sisk, were Albert, who married Mary (Polly) Richards Hall; Evaline, who married Austin Vinson; John, who was unmarried; and Calvin.

Thomas' children by his second wife, Lydia Davis, included Judge Edwin James Mantooth, who married Caledonia Watson; Lafayette, a physician who married Mary (Molly) Annetta Moore; Lucinda, who died young; W.B., who married Betty Haynes; Florence, who married W.H. Bonner; Hester Samantha, who married Benjamin Lloyd McPherson; Thomas Crittendon, who married Caledonia Nelson.

Thomas, Lydia and their daughter Lucinda died on the same day, July 28, 1865. According to a family story, they were accidentally poisoned by a doctor. A casket was also made for their youngest son, Thomas, and their bodies were held until his death. However, he recovered. Thomas' grandson, Albert Edwin Mantooth, said the deaths were probably the result of smallpox and often described how all the family's clothing was burned to kill the smallpox germs.

In Angelina County, Judge Edwin James Mantooth attended a private school and later returned to Tennessee to complete his education at Cumberland University, Lebanon, graduating in 1879.

That same year he was admitted to the Texas Bar and moved to Homer, the county seat of Angelina County. He was elected county attorney, but declined a second term and entered private practice.

In addition to his law practice, Judge Mantooth was a partner with W.H. Bonner in a general mercantile business at Homer. An early ad for the business in the March 2, 1883, edition of *The Homer Banner,* described the business: "Mantooth and Bonner, dealers in all kinds of drygoods and groceries. We make a specialty of gent's furnishings and goods, including ready-made clothing for men and boys. Don't fail to call on the above for whiskies, cigars, etc., for we keep the finest, both at Homer and Lufkin, and Mr. H.D. Weaver at Lufkin and Mr. Tom C. Mantooth at Homer, will take pleasure in waiting upon you at any time between 4 a.m. and 4 p.m., except Sunday."

When Lufkin was established on the Houston, East and West Texas Railroad in 1882, Mantooth, like other professional people and businessmen, moved from Homer to be near the railroad. He built a large home at the corner of Mantooth and Raguet Streets, the site of the present Mantooth Park.

When the City of Lufkin was incorporated in an election on October 15, 1890, Mantooth was proposed as one of the first aldermen, as well as the first city secretary. He joined Mayor J.M. Smith and aldermen H.D. Weaver, W.A. Abney Sr., J. Kerr, and J.F. Davis on the first city council, according to the results of a second election on November 15, 1890. On November 20, the aldermen called a special meeting at the offices of Mantooth and Townsend to accept their oaths of office and discuss plans for formalizing the city government. The city was incorporated with a property valuation of $250,000.

In 1892, Lufkin elected its second mayor, Thomas Crittendon Mantooth, a brother of alderman E.J. Mantooth. In 1916, when Thomas died, *The Lufkin Daily News* wrote: "Mr. T.C. Mantooth...died at his home this morning at six o'clock. This sad occurrence was not unexpected by the members of his family, nor his many relatives and friends. For the past several months, he has been suffering with an affliction which the most eminent specialist in the state pronounced incurable. The deceased was one of the most prominent citizens of this town and county, was born within one and a half miles of old Homer, and has been identified with the interest of this locality for many years. He was 53 years of age, leaves a wife and five children." At the time of his death, Mantooth was serving another term as mayor of Lufkin.

Judge E.J. Mantooth also served as Lufkin mayor from 1895 to 1897 and from 1901 to 1903.

In addition to this partnership with Dr. Alexander Madison Denman in bringing telephone service to Lufkin, Judge Mantooth served as president of Lufkin National Bank, as a director of Texas Power and Light Company, as a director of the Lufkin Ice Company, and the Angelina Hotel Company. He was also active in the Texas and American Bar Associations, the Chamber of Commerce, and in numerous political and civic organizations.

As an attorney, Judge Mantooth's most famous case was probably one that shocked, and literally, shook Lufkin on March 1, 1913, when the town's railroad depot was dynamited.

The explosion occurred on a Sunday night about 10 o'clock and resulted in hundreds of thousands of dollars in damage. Frank Parsons, a railroad employee, was the only man on duty at the time of the blast, but it was three years before Lufkin knew for certain if Parsons was alive or dead.

During those years, Parsons became a local legend. In the minds of many Lufkinites, there was little doubt he was dead. The debris of the bombed depot had yielded the remains of a body, and it was soon ruled the bones were those of an animal, but many Lufkinites felt Parsons went to his death faithfully on duty at the depot.

A high court agreed and after legal battle, Parsons' step-mother collected a hefty insurance settlement and Parsons was declared legally dead.

Lufkin's Railroad Depot explosion, March 1, 1913. Photo, Bob Bowman Collection.

But Judge Mantooth, representing the insurance companies, remained suspicious of Parsons and on January 4, 1916, a Western Union telegram from Mantooth in California to authorities in Lufkin sent a new shock wave through Lufkin: Mantooth was returning Parsons to Lufkin.

With the services of a private detective, Mantooth had found the key to the mystery: a letter had been sent to the clerk of a Galveston court seeking information about the case involving Parsons. Although he had used an alias, his writing gave him away. Mantooth was soon on the trail. Further investigation showed Parsons had enlisted in the Army since disappearing from Lufkin and was given an honorable discharge.

The Mantooth family from left to right, top row, Mollie Mantooth Simons, Winnie Mantooth Shotwell, Chloe Mantooth Collins, and Cleo Mantooth Denman. Front row, left to right, Robbie Mantooth Bowers, Mrs. E. J Mantooth, E. J. Mantooth and Dell Mantooth Lanter.

According to *The Lufkin Daily News* of January 6, 1916, when Mantooth confronted Parsons in California, he said: "I am glad you came. I am glad it is all over."

The newspaper headlined the Parsons capture with, "FRANK PARSONS ARRESTED IN CALIFORNIA," and explained: "There are those in Lufkin who never expected to see such a headline while there are others who have been looking for it for nearly three years. It is a matter upon which people have differed widely in opinion as it is possible for individuals to differ upon any subject. Even yet, after hearing last evening what is to follow in this item, there are to be found a few who are still skeptical and will remain so until their own vision sets all doubt aside."

Judge E. J. Mantooth

The newspaper on January 7, 1916, reported: "When Mr. Parsons stepped off the back platform and descended the steps on the opposite side of the depot, there was the unanimous expression from those who gazed on his features: 'That is him.'"

While Judge Mantooth's persistence and Parsons' return to Lufkin settled the mystery of his disappearance, the dynamiting of the depot was another matter. The wheels of justice cranked into motion, a trial was set for February 7, 1916, in Center, 60 miles northeast of Lufkin, and Mantooth and the insurance companies began their efforts to recover the money they paid to Parsons' step-mother.

Mrs. E. J. Mantooth

But a jury wrote the final chapter of one of Lufkin's most bizarre cases, and Judge Manooth's most celebrated legal effort. Parsons was tried, found not guilty, and freed. He was never seen in Lufkin again.

During their lives, Judge Mantooth and his wife, Caledonia, were devoted church goers.

They were converted to the Christian Church around 1884 in old Homer. When the congregation moved to Lufkin in 1894, it was located in a building known as the "Free School Building," located at the corner of East Grove and Culverhouse Streets. On September 7, 1903, the site was deeded by Mantooth to the trustees of the Christian Church of Lufkin. The building was later remodeled and made into a church and included three colored glass windows.

In addition to his civic enterprises, Judge Mantooth had extensive farming interests in Angelina County.

Mantooth and his wife had eleven children. They were:

- Dell, who married Will Lanter.

- Alberta Maude, who died when she was 11 months old.
- Thomas Winfred, who died when he was 15.
- Louie, who married Lula Bonner.
- Milton, who married Muriel Starr.
- Edwin Wood, who became a doctor and died young.
- Winnie, who married George Shotwell.
- Chloe, who married C.B. Collins.
- Cleo, who married Kester W. Denman.
- Robbie, who married Harry T. Bowers.
- Mollie, who married Shirley Simons. Simons was a renowned Tyler architect who designed several homes for the Mantooth and Denman families, as well as a number of public buildings and churches in the Lufkin area.

Alexander Graham Bell, the inventor of the telephone. Photo courtesy, Texas Telephone Museum.

4 The history of telephones in Texas

When Dr. Alexander Madison Denman and Judge Edwin James Mantooth strung Lufkin's first telephone line between their offices in 1898, the telephone was still a relative novelty in Texas.

Most historians credit Colonel A.H. Belo, publisher of the *Galveston News*, with the installation of Texas' first telephone on March 18, 1878. The line reportedly ran between Belo's newspaper office and his Galveston home. It was one of the first 1,000 lines installed nationwide.

Subsequent demand for telephones by other Galveston residents led to the creation of a switchboard to switch calls from one line to another, and the first exchange was opened for business at Galveston on August 21, 1879. Another exchange was opened in Houston shortly thereafter.

The first exchanges were built by the Western Union Telegraph Company under patents awarded to Thomas A. Edison. According to present standards, early switchboards were quite crude. Calls were made by battery signals operated by push buttons, and almost all lines were party lines. The first type of magneto signal bells were long narrow boxes from which a large button projected in front. By striking the button with the hand almost hard enough to crack the plaster on the wall, a current from the magnet inside signaled the operator in the central office.

The Southwestern Telegraph and Telephone Company, which was organized in 1881 to operate exchanges in Arkansas and Texas, took over the Galveston and Houston exchanges and started exchanges at Austin and San Antonio.

A Dallas exchange with forty subscribers opened on June 1, 1881, and a Fort Worth exchange opened on September 1 of the same year with about the same number of subscribers.

An exchange opened at Waco October 1 with forty-five subscribers and during 1882 exchanges were opened in Brownsville, Brenham, Cleburne, Colorado, Corsicana, Gainesville, Greenville, Jefferson, Marshall, Paris, Palestine, Sherman, Denison, and Texarkana.

The next step in development for the new Texas industry, connecting individual exchanges by long distance lines to allow people in one town to talk with those in other towns, was completed in 1883 with construction of the state's first long distance line linking Galveston and Houston.

In 1883, eastern capitalists purchased Southwestern Telegraph and Telephone Company, and established temporary headquarters in Little Rock, Arkansas, followed by offices at Austin. A little more than a month later there were 2,344 subscribers statewide.

Galveston, the largest exchange, had 330 subscribers; San Antonio, 323; Dallas, 259; Fort Worth, 168; and Houston, 148. The following year, in an effort to improve unreliable service, the company established a precedent when it

An early Texas telephone. Photo, Texas Telephone Museum.

introduced women as operators during night hours at Galveston. Boys had initially been used as operators, but their rude remarks to customers and their erratic operating practices were a trial to telephone users. The practice of using women as operators was soon established in all exchanges.

While sources disagree, it is generally accepted that the first PBX board (private branch exchange switchboard) in the United States was installed in the Capitol Hotel, later renamed the Rice, at Houston in 1882.

The first common battery switchboard, where telephones operated from energy supplied by centrally-located batteries instead of individual batteries at each telephone, was installed in Dallas on February 13, 1899, when the Dallas exchange had 1,200 subscribers. After 1900 telephone systems in Texas continued to grow as equipment improved, long distance lines were extended, and more telephones were added, reaching 102,443 telephones statewide.

The first automatic telephone exchange in Texas was built at Cleburne in 1903, but stopped operations in 1912, when a fire destroyed much of the equipment. The first dial exchange of the Southwestern Bell Telephone Company was installed in Dallas on July 30, 1921, and when the Wichita Falls exchange was changed to automatic operation on December 10, 1921, it was the first city to have all dial telephones.

The year 1912 also marked the start of an era of consolidation in the telephone industry. That year, the Missouri and Kansas Telephone Company, the Missouri Bell Telephone Company, the Pioneer Telephone Company of Oklahoma, and Southwestern Telegraph and Telephone Company were consolidated to form a new operating unit known as the Southwestern Bell Telephone Company with headquarters in St. Louis.

By the end of 1946, the company operated more than 930,000 telephones in 247 exchanges, while 349 other Texas telephone companies with 794 exchanges operated a total of 182,704 telephones.

The largest telephone centers, all Southwestern Bell exchanges, were Houston, with 169,068 telephones; Dallas, 147,497; San Antonio, 85,323; Fort Worth, 73,566; and Austin, 34,030. Southwestern Bell had 20,000 employees in Texas at the beginning of 1947 and handled an average of 7,165,000 local and 130,000 long distance calls every day.

Telephone service in Texas increased rapidly after World War II, following an era when the war restricted supplies required for telephone services. By 1955 there were 311 telephone companies operating 2,398,521 telephones statewide, and Texas ranked seventh in the nation in number of telephones (all figures here include main

Colonel A. H. Belo. Photo, Texas Telephone Museum.

line telephones and do not include extension telephones).

On May 7, 1955, the state's two-millionth telephone was installed in the governor's office at Austin. Southwestern Bell Telephone Company inaugurated direct distance dialing in Harlingen in 1955, and by 1963 Texas had more than 550,000 telephones on direct distance dialing networks.

As telephone systems were consolidated, the Texas industry grew from 6,266,000 telephones operated by 108 companies on January 1, 1971, to 7,058,993 telephones served by ninety-eight companies on December 31, 1972.

In that year Texans served by Southwestern Bell Telephone Company, the largest company in the state, made a daily average of 35,221,822 local and long distance calls. The city of Houston, with 1,369,069 telephones, had the largest number of any metropolitan area in Texas. By January 1, 1975, there were 7,747,100 telephones in the state, more than 90 percent of which were on direct distance dialing.

Of the ninety-six companies then in operation, ninety were members of the Texas Telephone Association, including Southwestern Bell Telephone Company and Mountain Bell Telephone Company in El Paso County.

By 1980, Texas telephones had increased to more than eleven million, served by eighty local exchange companies, including 78 independents. Among the independents were 23 telephone cooperatives owned by subscribers, serving more than 105,000 telephones through 217 exchanges state-wide. Bell System companies continued to serve four out of five telephones in the state, while independent companies served over half the state's 250,000 square miles of certified service territory.

Southwestern Bell Corporation became a separate entity when the Bell System divested in 1984 and was thenceforth no longer associated with AT&T. The largest subsidiary of the new company was the Southwestern Bell Telephone Company, which provided service to nine million customers in five states, including Texas.

Also established in 1984, the Texas Division of Southwestern Bell Telephone, with headquarters at Dallas, became one of the largest nongovernmental employers in the state, serving 6.2 million customers in 549 Texas communities.

By 1990, the Lufkin-Conroe Telephone Exchange, with 63,717 lines, was one of the major independent telephone companies. Others included GTE Southwest (1,204,791 lines), Contel of Texas (167,801 lines), Central Telephone Company of Texas (127,393 lines), and United Telephone Company of Texas (110,750 lines).

Because customers now own the wiring within their premises as well as inside equipment, the industry no longer counts total telephones served. Access lines reflect the number of connections provided by the company, but do not necessarily equate to number of customers.

A "candlestick" telephone of the type used in the early 1890's. Photo courtesy, Texas Telephone Museum.

In 1996, 58 independent telephone companies in the state included twenty-four telephone cooperatives serving more than 100,850 access lines through 204 exchanges while Bell System companies served roughly 80 percent of all Texans.

Angelina County farmers hauled their cotton to Lufkin's Cotton Square for shipment on the HE&WT Railroad. Photo, Bob Bowman collection.

5 Lufkin at the turn of the century

Lufkin around the turn of the century was a rough-hewn settlement where life revolved around the railroad, sawmills, and saloons.

Conductors on the Houston, East and West Texas Railroad often admonished passengers to remain on the train when it stopped at Lufkin and at least one conductor regularly called out the stop with a warning, "Next stop, Lufkin. Prepare to meet thy God." Lufkinites often wandered down to the depot when the trains arrived to pick fights with passengers who dared to comment on the town's rough-looking appearance.

In 1898, when it received its first telephone, Lufkin was only 16 years old, having been founded by the HE&WT in 1882. The community was incorporated as a city in 1890 and became the Angelina County seat of government in 1892.

Lufkin's leadership, which included telephone pioneers Alexander Madison Denman and Edwin James Mantooth, had sought to make Lufkin the county seat since the early 1890s. Most of Homer's businesses had moved to Lufkin, the doctors and attorneys had abandoned Homer, and Lufkin was attracting new residents every week.

In 1885, an attempt was made to move the county seat to Lufkin in an election on May 9. Homer, however, won the balloting, 532-247.

But with a majority of the county's population concentrated within its boundaries, Lufkin continued to argue that it should be the county seat. Homer remained adamant. The courthouse had been there 30 years; it would remain there.

In November, 1891, the people of Homer greeted the dawn and found its courthouse in ashes. The town was certain the fire had been deliberate, the result of orders originating in Lufkin and carried out by night riders, but it lacked proof.

Angelina County was named for Angelina, an Indian maiden immortalized by this statue in Lufkin. Photo, Jimmy Bowman.

Everything was destroyed in the fire except two fireproof safes containing the county's records and money. The safes were so hot that it was three days before they could be safely opened. During the time, Homer residents saw that they were closely guarded to prevent possible theft by Lufkinites.

On November 30, 1891, the Commissioners Court met to designate temporary offices for the county. The back room of Dr. W.W. Manning's drug store was to be used by the county clerk

An early locomotive of the type found in Lufkin in the early 1900's. Photo, Bob Bowman Collection.

for a rent of $4 a month and "the house on the north side of the public square in Homer, known as the Old Collins Hotel," was set aside as a temporary courthouse.

Lufkin wasted no time in capitalizing on Homer's loss. The day after the fire, the Commissioners Court received a petition from W.B. Mantooth, a brother of Judge E.J. Mantooth, and 200 other Lufkin residents for an election to decide if the courthouse would go to Lufkin or remain in Homer.

Twenty Lufkinites agreed to furnish land and a new courthouse if the county seat was relocated. They were Judge E.J. Mantooth, Dr. A.M. Denman, L. Mantooth, A. Chancey, W.J. Townsend, S. Abram, I.D. Clark, J.A. Abney, A.J. Smith, J.E. McPherson, W.M. Jones, A.W. Ellis, J.M. Scurlock, J.F. Davis, W.G. Chancey, J.P. Eddings, William Chancey, T.P. West, Sam Berry, J.F. Weeks, A.C. Vinson, and H.D. Weaver.

When the election was held on January 2, 1892, Lufkin totaled 2,086 votes and Homer received only 436. Homer residents argued, however, that Lufkin had brought in dozens of railroad workers, traveling salesmen, and other out-of-towners. It was a bitter loss for the community which once exerted so much leadership in the county.

On January 9, meeting at Lufkin for the first time, the Commissioners Court ordered that the "lower story or room of the Masonic hall in Lufkin be used as a courthouse" until permanent quarters could be found. About a month later, the Commissioners Court sold the "nails and other rubbish in the ruins of the courthouse burned at Homer."

The Houston, East and West Texas Railroad deeded one half of Block 15 in Lufkin for $150 for a courthouse square and on February 16, 1892, the Commissioners Court outlined specifications for the Lufkin courthouse, a wooden frame structure, to occupy the square. Two dozen Lufkin residents, including the Denman and Mantooth families, pledged a total of $2,000 for construction of the building. It was completed in December of 1892.

By 1898, Lufkin's population was well above a thousand people. The town had two churches, a small college, a public school, a bank, a sawmill, a brickyard, several cotton gins, a weekly newspaper, a lumberyard, a tannery and other conveniences. The town also had more than 100

Cotton farmers on the shipping deck of the HE&WT Railroad in Lufkin. Photo, Bob Bowman Collection.

businesses and professional people.

The introduction of the telephone by Dr. Denman and Judge Mantooth in 1898 opened up new communications possibilities for the growing community.

On June 12, 1899, at a meeting of the Lufkin City Council, with Mayor W.M. Glenn presiding, the council granted a franchise to the American Telephone and Telegraph Company and its subsidiary, Southwestern Telegraph and Telephone Company, to run its lines "through and over the streets, alleys, creeks and bridges" of the city.

The company was given "the right to erect and maintain poles, wires, machinery, apparatus and appurtenances necessary for telephone and telegraph purposes," but the City Council reserved the right to designate where the poles and wire should be placed.

In exchange for the franchise, the Council required that the mayor be furnished one telephone and that the instrument be maintained as long as an exchange was maintained in the town. The Council also stipulated that the city should be given "sufficient space upon the top cross arm" of each telephone pole for its municipal fire alarm telegraph system.

While the Denman/Mantooth telephone system provided local communications, the AT&T system interconnected with the Denman/ Mantooth switchboard to provide long distance service to Lufkin. To do this, AT&T invested in switching equipment and copper lines to the outside world and provided operators in Lufkin. The city franchise was necessary for the company to conduct business in Lufkin and for the use of the streets and alleys instead of formal rights-of-way.

Years later, this arrangement changed and the Lufkin company began to own the long distance service for the Lufkin area. A compensation arrangement was developed to share the costs and revenues from long distance. At that time, several AT&T employees became Lufkin employees.

In his diary, Mayor Glenn described the arrival of the long distance telephone system: "The system was...ready for operation in August of that year (1899). This first system was fine, too. When you put the receiver to your ear, you could hear nearly everybody talking at the same time."

Veterans of the Civil War gathered in front of the building housing the Lufkin Telephone Company around 1900. The telephone operations were housed on the second floor above the drug store. Photo courtesy, Museum of East Texas.

Lufkin's first telephone exchange was housed on the second floor of the building at far right in this 1909 photo taken by a Lufkin photographer. The lower floor housed a drug store. Photo courtesy, Museum of East Texas.

6 Creation of the Lufkin exchange

By 1900, Lufkin businessmen were so impressed by the capabilities of the private telephone system of Dr. Alexander Madison Denman and Judge Edwin James Mantooth that they began to clamor for telephones of their own.

In response, Dr. Denman, Judge Mantooth and a third investor, Eli Wiener, invested $500 apiece and opened a small, one-room telephone office, installing a magneto switchboard using "crank" energy.

Retired banker H.J. (Bubba) Shands, Jr., says his grandfather, W.J. Townsend, a partner with Judge Mantooth in the law firm of Mantooth & Townsend, was invited to invest in the telephone company for $500, but declined. "Grandpa told Judge Mantooth that he didn't think the fool thing would work," said Shands.

The early switchboard installed by Denman, Mantooth and Wiener had a capacity of about 800 lines, but only 40 lines were installed initially to serve about 100 telephones in the city. Most were installed in businesses and in the homes of affluent residents. The company had only four employees.

A later article in the *Stromberg-Carlson Transmitter*, a company publication, said Mrs.D.C. Hucherson made the first call over the magneto switchboard to Peyton R. Denman.

The company's first office was in a single room on the second floor of the Denman Building, a downtown structure built around the turn of the century by Dr. Denman and his family. Located at the intersection of Dozier (today's Frank Avenue) and First streets, the building housed a number of businesses, including a drug store and a cafe. The building was remodeled in 1939, according to Dr. Peyton L. Denman, a grandson of the builder, which accounts for the 1939 date on the building's facade. The building exchanged ownership several times since 1939 and was purchased by the telephone company in the 1980s.

Dr. A. M Denman served as Lufkin mayor from 1903-05. Photo courtesy, Museum of East Texas.

Eli Wiener, who joined Denman and Mantooth in the fledging company, was a lumberman and part owner of Angelina County Lumber Company, as well as one of the founders of Lufkin Foundry and Machine Company in 1902. He later sold his interest in the telephone company to Denman and Mantooth and moved on to other financial interests.

May 4, 1901 fire in downtown Lufkin. Photo, Bob Bowman Collection.

While growth of the new local system was steady the first year, it received a setback in May of 1901 when a massive fire swept through the downtown sector and the residental area immediately north of the business district. Starting in a cafe on Cotton Square, the fire consumed 40 homes and businesses, including several with telephones.

A second fire in August destroyed the Crutchfield home, portions of Lufkin National Bank, where Judge Mantooth and Townsend had their law office, and damaged several other businesses.

Responding to the two fires, Mayor W.M. Glenn and the Lufkin City Council passed an ordinance prohibiting the construction of wooden frame buildings in the downtown business district. Lufkin began to rebuild after the fires and growth resumed among its businesses, including the new telephone company.

The Lufkin Telephone Company's switchboard brought to Lufkin a new tradition, "central's operator," a group of women who began changing the communications, social activities and business practices of Lufkin.

The company's switchboard operators, working side by side in the small room in the Denman Building, seemingly knew everything about everybody. They knew when women went shopping, when their children came home from school, and when business appointments were made. They were also privy to many of the vices, weaknesses, and suspicions of Lufkin, but they were sworn to the decades-old tradition of switchboard operators: never, never spread gossip.

Dr. A. M. Denman posed for this early 1900's photo in front of his home. Photo courtesy, Kurth Memorial Library.

The people of Lufkin came to rely on the phone company operators for information they couldn't get anywhere else. If someone wanted a doctor, they often called "central," or if someone wanted to leave word with someone else about an event, they sometimes used the operator as an earlyday voice mail.

Central's operators were readily identified by their voices and were gifted with the ability to memorize the names of hundreds of subscribers and their telephone numbers. While each subscriber had a telephone book—a one-sheet list of numbers in the early years—most called "central" for a number or told the operator, "Miss Emma, can you get me Winnie Bell?"

People who lived in that era would, years later, remember it as a time when

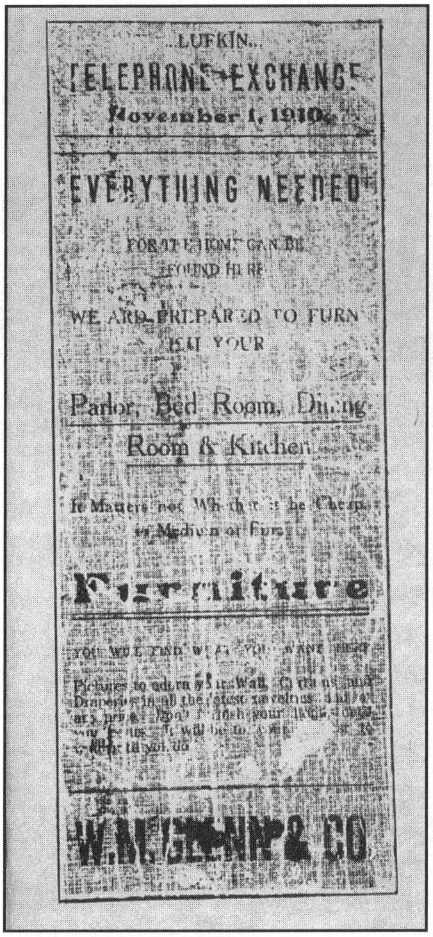
LUFKIN
TELEPHONE EXCHANGE
November 1, 1910.
EVERYTHING NEEDED
Parlor, Bed Room, Dining
Room & Kitchen
Furniture

1910 Lufkin Telephone Exchange Directory cover.

people grew closer, relied on the telephone as a cherished communictions instrument, and relished their conversations "down the line."

On February 26, 1908, the Lufkin Telephone Company entered into an agreement with the Southwestern Telegraph and Telephone Company to attach its telephone wires to poles owned by Southwestern in Lufkin. The agreement covered 825 pole contacts and approximately 22 and 1/2 miles with the Lufkin company paying $91.66 a year for the connection rights. Southwestern and the Lufkin Telephone Company soon began sharing space on the second floor of the Denman Building, each operating a switchboard in the same room.

On October 1, 1908, Dr. Denman, who had served as president of the company since its inception, suffered his fatal automobile accident. Judge Mantooth became president of the company and Peyton R. Denman, Dr. Denman's son, became the secretary and general manager. In 1910, Dr. Denman's heirs and Judge Mantooth acquired Weiner's interest in the telephone company.

The company was also incorporated in 1910 with a capital of $10,000 and the same year the Lufkin City Council gave the exchange the "right, privilege and authority to construct, operate and maintain upon the streets, alleys and public ways of the City of Lufkin poles, wires and fixtures necessary for the purpose of furnishing to the citizens of Lufkin and the public generally, as well as corporations or compacts of individuals, telephones, telephone poles and wires, and other appurtenances, devices, machinery, etc., for such use of said telephone plant...for a period of 25 years."

Road construction by mule teams in the early 1900's. Photo, Bob Bowman Collection.

Lufkin Avenue, looking east from Cotton Square in 1909. Photo, Bob Bowman Collection.

As a part of the franchise, the exchange agreed to furnish to the city 10 telephones for the life of the franchise for the rent of $2.50 per month for each telephone. Judge E.J. Mantooth, as president of the Exchange, and secretary and general manager Peyton R. Denman, Dr. Denman's son, signed the franchise for the telephone company.

The 1910 Lufkin telephone book offered these instructions for using the telephone:

"The person calling can obtain the best results: By always looking up the number in the latest directory, thereby avoiding error calls. By speaking distinctly into the transmitter. By being sure that the operator has understood and repeated the correct number. By being patient and allowing the party called a reasonable time to answer..."

The 1910 telephone book included, among others, these numbers in numerical order: 1. H.A. Fuller's residence. 2. S. Abram's residence. 3. W.J. Townsend Sr.'s residence. 4. Lewie Markus's residence. 5. Dr. J.C. Van Nuys' residence. 6. Mrs. Mat Campbell's residence. 7. Purvis Meat Market. 8. L.I. Frank's residence. 9. Angelina County Trading Company. 10. W.A. Abney's residence. 11 and 12. D.A. Singleton's residence and store. 13. Fred J. Luck's residence. 14. Angelina County National Bank. 15. Lufkin Land and Lumber Company office. 16. R.A. Knight's residence. 17. Lufkin Land and Lumber Company store. 18. Murphey Hotel. 19. Dr. T.W. Largent's residence. 20. Rev. W.H. Crum's residence.

Correspondence in the company's archives reflects a flurry of activities in 1911 to bring about improvements in the telephone service for Lufkin and the surrounding area.

On January 26, C.E. Walton, a Huntington, Texas, merchant, wrote the telephone exchange a letter complaining about poles the company had apparently picked up at his house. "...these poles are worth 75 cents each to me as the boys got them right at my house I am building. I could have used them so handy and was intending to use them as sills in stalls. They did not ask me anything about them; they got them without permission. If they had asked me, I would have sold them poles cheaper, but not at that place. I would not have sold them at no price. Eight poles, 75 cents each, $6.00. Please remit..."

May's records reflect a series of checks paid for various exchange activities, including $20.00 to Mrs. A.M. Denman (the widow of Dr. Denman) for a month's rent (apparently for the company's office), a $22.32 printing bill to Dorsey Printing Company, a $15.70 check to the Lufkin Electric Light Company for power, $2.70 to Lufkin Foundry and Machine Company, and apparent payroll checks to Mabel Moore, C.H. Harrell, H. Beauchamp, Clarence Madden, Willie Perry, Stella Cogdell, Winnie Green, George Shotwell, and W.E. Bynum.

On July 22, W.C. Trout, secretary and manager of Lufkin Foundry and Machine Company, wrote E.G. Pike, general manager of the Southwestern Telegraph and Telephone Company at Houston, this caustic letter: "The service we get through the Lufkin exchange is anything but satisfactory to us. Most of our business is long distance calls and we have our own telephone exchange and hire a girl here to look after it for us, but it is almost impossible to get a satisfactory long distance conversation without the local central cutting us off or butting in some way. We write to ask if the business you get from this institution would not warrant you putting in a direct wire to your Southwestern exchange. It would mean more business for you, give us better service. We are not more than 3,000 feet away from the Central exchange and you can come in on poles that are already up and there will be no expense except the wire. If it pays you to have booths in the two hotels at Lufkin, it will certainly pay you to give us direct connection as well."

On August 18, Pike wrote E.J. Mantooth: "The Lufkin Foundry and Machine Company wants us to place a telephone in their office connected directly to our switchboard at Lufkin for the purpose of furnishing long distance service. Will you kindly advise whether or not this arrangement would meet with your approval. The telephone would simply be placed to facilitate handling long distance business, and if agreeable, we would probably install the instrument." Whether the agreement was ever reached is not reflected in the archives.

A series of letters in September of 1911 reflect efforts by the exchange to

A downtown Lufkin street scene in the early 1900's. Photo, Bob Bowman Collection.

The famous Hoo-Hoo Band of Lufkin in 1904. Photo, Bob Bowman Collection.

expand its services to Clawson, using the right-of-way of the Eastern Texas Railway. On September 1, a letter to W.F. Feagin, general manager of the Eastern Texas, from an unsigned official of the telephone company reported that Dr. Denman, before his death, "made some kind of contract for the right-of-way of this phone line over the Eastern Texas Railroad to the station at Clawson, and this telephone line was extended to Clawson Mill (a sawmill)."

Another letter to William N. Neff, president of the railroad, also written on September 1, by an unnamed telephone company official reported: "I have received your telegram regarding the right-of-way...and I have been trying to locate Mr. W.F. Feagin, who made the contract with Dr. A.M. Denman...but I have not been able to locate him..."

Still another letter, also written in September, went to John A. Sargent of Kansas City, asking about Feagin's whereabouts.

On September 5, Sargent, traffic manager of the Central Coal & Coke Company of Kansas City, wrote that Feagin was connected with the Kansas City Terminal Railway Company. Sargent promised to contact Feagin.

On September 8, the telephone company advised Neff that Feagin had been located and that efforts to finalize the right-of-way issue were nearing an end.

Before the end of 1911, Mantooth—who apparently felt the telephone company needed a full-time manager—persuaded the board to hire his son-in-law, George Shotwell, who had been running a Lufkin saloon, as the manager. Shotwell held the job until his death in 1923.

By 1912, the telephone system was doing so well in Lufkin that an unidentified competitor was considering installing another local system in the community.

In a letter to Judge Mantooth on March 28, an official of the Western Electric Company of Dallas said he was "sorry to learn that there are prospects for a dual system in Lufkin because, from a financial view, I do not think that the town is large enough to support them and allow either company a just return on the investment."

On June 17, Mantooth was sent a letter from A.J. Peavy, president of the Peavy-Byrnes Lumber Company of Shreveport, apparently in reply to Mantooth's letter of June 15. "I have your letter of the 15th and note same carefully, and in replying to say that I am somewhat surprised that such a report would get out that I was taking stock in the new telephone company, which I assure you is a mistake." Peavy said no one in the Peavy family had subscribed for any of the stock and added: "I feel that... the town is not large enough for two telephone companies, and do not feel that you have any occasion to worry..."

Mantooth apparently felt the competition might come from El Paso because on September 10, in response to a Mantooth letter of September 4, the El Paso law firm of Turney and Burges informed the Lufkin exchange "there is no Southwestern Telegraph and Telephone Company in El Paso." The firm said "the Automatic Company (apparently a telephone company) went out of business here some time ago...as a financial proposition, it was an absolute failure."

Mantooth wrote Turney and Burges on September 13, saying "the Automatic is undertaking to install a plant here, that is, they are attempting to sell stock for the purpose of putting in a plant here, and they are required to sell as much as 49% to citizens of the town, and up to this time, they have not been able to sell any stock." Mantooth felt the opposition might dismantle its El Paso plant and move it to Lufkin.

Mantooth received another letter from D.W. Williams of the Southwestern Telegraph and Telephone Company of Cleburne. The letter, dated September 12, talked about a possible opposition company in these words: "I do not feel like making you a clear answer as to the disadvantages of the Automatic system, on account of being in charge of an exchange in opposition to their system, but I will say that if I had to choose between the Automatic and the old fashion system that works by turning a crank to get the operator, I would choose the latter for my part."

The Lufkin exchange began considering measures in 1912 to improve its service by upgrading its equipment and lines. In a March 29 letter to Peyton R. Denman, who was living in Houston, the Western Electric Company enclosed specifications and a cost proposal for a new central battery office switchboard with supplementary equipment. The company said the $4,360 system's battery would carry "the present equipment of 500 subscriber lines for 48 hours or the ultimate equipment of 1,000 lines for a period of 24 hours without recharging."

On June 25, the Lufkin company provided statistics from the Lufkin operation to George E. Dutton of the Southwestern Telegraph and Telephone Company of Dallas, Lufkin's long-distance supplier. The charges for telephones below the toll lines ranged from $2.50 to $30 a month. "There are mills on these toll lines where we have telephones and charge them from $4 to $30 per month flat rate." The letter said some of the lines ran "two, three, four and five miles out of Lufkin."

In April of 1913, a dozen Lufkin residents—J.V. Lankford, Ben F. Cox, C.M. Maberry, G.C. Perkins, J.N. Brevelle, Hosa St. Clair, B.S. Brazil, Arthur Davidson, T.J. Maberry, C.A. Hill, J.L. Perkins and I.R. Cox—agreed to build a telephone line at their own expense to connect with the Lufkin exchange. The company agreed to connect the line with its system and to provide service to the 12 families for the rate of $6 a month.

Correspondence in 1915 reflects the details of a controversy that arose over the death of J.T. Nicholson's hog by a telephone worker.

Nicholson wrote to George Shotwell of the company on June 24: "I was in town two days ago to see you about some of your men running over one of my hogs, and killed it. I was talking to some of your linesmen and they said you would settle with me. I told them that the hog was worth eight dollars, but if you want to do what's right, I will square off with you for five dollars. If you will send me a check for five dollars, I will appreciate it very much. Let me hear from you at once."

Kester W. Denman, Sr., an attorney for the telephone company, replied to Nicholson on June 26: "I would appreciate a more definite and detailed statement concerning the matter...the date of the accident, where it occurred, who caused same, and in what manner by what means, the weight of the hog, its age, etc." Denman added: "If you are in need of money at this time, I will give you $2.50 without further investigation."

Nicholson replied to Denman on July 1: "...regards that hog you want me to tell you more about, the hog was killed between Mr. Summer's and Mr. Ballard's. The hog was hit with the heavy side of the car... on the 8th of June. The hog would have weighed between 75 and 100 pounds, its age was near 8 months. The hog was really worth $8... but as I already said I would square it off for five dollars. This is less than I said I would take but if you are willing to do what is right, please send me the amount of five dollars, and I am sure you are willing to do fair about it."

The telephone company apparently settled with Nicholson for the five dollars. On July 8, he signed a settlement paper with the company for that amount.

By January of 1917, the Lufkin telephone exchange had grown considerably and had 611 subscribers, 102 miles of pole lines, 630 miles of wire, and one

The Angelina County Courthouse in Lufkin built in 1903, was a landmark before it was demolished in the 1950's. Photo, Bob Bowman Collection.

central office, according to a report filed with the Interstate Commerce Commission for the previous year. The exchange was connected with Southwestern Telephone and Telegraph Company for long distance service. The company was also connected with telephone exchanges in Huntington, Broaddus, Chireno and Apple Springs and had operating revenues of $18,260.40, including $427.20 from long distance service. Operating expenses totaled $10,457.46.

A company balance sheet in 1918 showed assets of $27,766.30, including $18,940.80 in poles, lines and telephones; $4,510.34 in cable; $2,030.00 in a switchboard; $487.86 in office fixtures; $196.00 in supplies; and $1,601.30 in cash. Equity included $10,000 in corporate stock, $16,000 in surplus; and $1,601.30 in undivided profits.

The company also dealt with a problem of someone using a telegraph company's lines in San Augustine County. In a letter on September 6 to Frank Tremble, Superintendent of Telegraph, a company official wrote: "We have your letter...relative to the use of your telegraph by some telephone company between Broaddus and White City...we do not know what telephone company could be using your line. We are not. We have our own circuit from Lufkin to White City along the line of the railroad on the opposite side from your line. There has been no occasion for us to use your line. There is a telephone company at Broaddus and it is possible that you will find the use there..."

In May of 1919, the Lufkin exchange entered into a new agreement with Southwestern for long distance service from Lufkin. The contract extended to April 1, 1920. The company also sought in September to acquire or lease a line of poles six or eight miles in length from Broaddus to Sturgis and made an application with the St. Louis and Southwestern Railway Company to use its motor car on the railroad's line between Lufkin and Ratcliff and between Lufkin and White City. In exchange the company agreed to give the railroad free local telephone rights "into each and every station on the line from Lufkin to White City and from Lufkin to Ratcliff and Kennard."

Judge Edwin James Mantooth ran the Lufkin Telephone Exchange from his law office on the second floor of Lufkin National Bank at the corner of Lufkin Avenue and First Street. Photo, Bob Bowman Collection.

7

The twenties: Lufkin's telephone service grows

By the 1920's, Lufkin's population was approaching 4,000 and, despite a lingering reputation as a town where liquor flowed freely and fist fights were still common on Cotton Square, the town was beginning to shape a progressive reputation.

The town's downtown mud streets were gradually disappearing under paving, the result of an aggressive program by the city government. The town had acquired its first piece of motorized fire-fighting equipment, a Model T Ford chemical truck, in 1914; the city's first city hall was located in a second floor above the fire station on Cotton Square in 1915; and a 15-man charter commission headed by Kester W. Denman, a son of telephone pioneer Dr. Alexander Madison Denman, completed its work in 1918 on a new municipal charter.

Despite the progress, Lufkin was struggling in the early twenties. Many of the town's most promising young men had died during World War I and the lumber industry, which had brought new jobs and payrolls to the community, was declining as a result of overcutting in the forests. Sawmill ghost towns were sprouting like spring wildflowers all over East Texas.

Lufkin Telephone Exchange, however, was growing. From an initial 100 telephone subscribers in 1900, the company entered the twenties with nearly 700 subscribers, 119 miles of pole lines, 820 miles of wire, and one central station.

The company's day-to-day operations remained in a second floor offices at the corner of Dozier (now Frank Avenue) and First streets, but much of the company's corporate activities occurred in the law offices of Judge Edwin James Mantooth and Kester W. Denman, Sr., located in the Lufkin National Bank building a block away at the corner of Lufkin Avenue and First Street. Judge Mantooth was president of the company from 1920 through 1928 and Mrs. Alexander Madison Denman was vice-president.

On May 15, 1920, the company issued this public notice: "The information has come to us that the public generally in the neighborhood of telephone No. 127 in the home of... has been using said telephone. It is an inflexible rule of this company to prohibit the use of one person's telephone by others. Therefore, (this) will be notice to all persons other than the family of Mr. ...to cease using said telephone."

In November, the company exchanged a series of letters with Lufkin Land and Lumber Company, a sawmill on Lufkin's east side. The general manager of the lumber company wrote George Shotwell, general manager of the exchange, on September 10: "We have your telephone bill...covering phone rent at Camp Bannister (in San Augustine County). And note again that you charged us $75.00 for the month's service. This service, as we have previously advised you, is not satisfactory for the reason we are continually being bothered with other people on the line...we cannot afford to pay $75.00 per month unless the line to the camp is working independent from other lines so that we can talk without being bothered by other parties and to use the line whenever we desire to talk to our camp..."

Shotwell received another letter from the sawmill on November 22: "We have your (letter) of the 16th, with reference to our new telephone line to Camp Bannister...you state that our trouble lies at Camp Bannister, but in this I believe you are mistaken, as we are having no trouble with anything except continual cross talk. At times, we are hardly able to talk to our man at Camp Bannister for the reason that the lines are either crossed or that someone has cut in..and we only ask that you make it a point to get on the line most any time during the day up to 9 o'clock at night, and you can ascertain for yourself what the real trouble is..."

Shotwell got a third letter from the sawmill on November 27: "With further reference to our letter of November 22 regarding (the) phone line to Camp Bannister, we wish to ask if you have talked over this line as suggested...on the 22nd. If so, advise us why it is you can hear everybody talking and hear all the phones ringing, etc. The talk and ringing is so continuous at times that we are unable to talk and hear over the phone, so therefore we are under the impression that somebody has cut in on this line, or else the line is crossed with some other line.."

In 1923, the telephone increased its capital stock from $10,000 to $75,000 and during the year named Jean Shotwell to succeed his father, George Shotwell, as general manager. George, who had served in the role since 1911, died in 1923. Jean Shotwell had also joined the company in 1911, working first as a collector and then as a lineman.

Convinced that Lufkin had a bright future, the telephone exchange embarked on a far-reaching expansion by replacing its 800-line magneto board with a common-battery Stromberg-Carlson "super service" facility with a capacity of 2,400 lines. The expanded facilities included the main switchboard, a cable and relay rack, power equipment, machine equipment, a main battery, a booster battery, and a noise-killer battery.

In a letter to Mantooth on July 24, 1923, a Stromberg-Carlson official apologized for a delay in sending contracts to Lufkin for the equipment, "this being vacation time (and) the Engineering Department was short of help."

In a contract signed later in July, E.J. Mantooth, president of the Lufkin exchange, and Kester W. Denman, Jr., secretary, agreed to pay the Stromberg Company $21,046, payable with $6,000 on April 1, 1924, and a series of 10 notes of $1,500 each from July 1, 1924, to October 1, 1926. Stromberg gave the Lufkin company an allowance of $1,500 for its old switchboard.

Lufkin's fire station and city hall stood at the end of muddy Lufkin Avenue in the early 1900's. Photo, Bob Bowman Collection.

To support the new equipment, Mantooth began discussions with the owners of the old freight department of the Eastern Texas Railway depot in Lufkin to rent the building, apparently as a temporary maintenance facility. "We have had an examination made of the freight department, and it is in quite bad condition, but (I) am advised that it can be used for the purpose desired as there would be nothing but some hardware material placed in the freight department," he wrote E.B. Perkins of Dallas on May 29. An agreement to rent the building for

12 months for $10 a month was finalized in June.

On the night of June 24, 1924, at 11 p.m., the switchboard was changed from the magneto type to a modern common battery system. Within two years, 1,100 lines were working and the company was handling 10,000 calls per day. Installation of the new equipment was in the second floor offices of the Denman Building.

A new fire station built in 1929 served Lufkin from Lufkin Avenue south of the downtown area. Photo, Bob Bowman Collection.

By the middle of May, before the Stromberg-Carlson equipment was purchased, C&I Engineering Company estimated the length of new pole lines to be located in the city, apparently to serve the new lines. The estimate included 6.43 new miles of lines with the greatest number of new lines occurring on Dozier Avenue, Bonner Street, Iron Bridge Road, near Lufkin Land and Lumber Company, on Jefferson Avenue, Abney Avenue, and on North First Street.

In May of 1923, Mayor Sam Kerr and the Lufkin City Council approved a new 25-year franchise for the telephone exchange. In return, the company gave the city free use of its poles and posts for fire alarm and police telephone boxes.

Also in May, the telephone exchange made an effort to collect a series of unpaid telephone bills totaling $791.75. The list included some of Lufkin's most prominent people, including the mayor.

In August, the telephone exchange signed a new agreement with Southwestern Bell Telephone Company to provide long-distance service to Lufkin. Southwestern's long-distance switchboard was again located in the Denman Building and agreed to pay the Lufkin exchange $800 a year.

In the summer of 1923, there was another flurry of correspondence between the telephone company regarding the logging camp at Bannister in San Augustine

Crowds of Lufkinites jammed Cotton Square in the 1920's for an event. Photo, Bob Bowman Collection.

County. On June 30, San Augustine County Lumber Company wrote the telephone company, "Referring to your June 1st bill amounting to $35.50 for rentals at new camp, San Augustine. In view of telephone rent being paid in advance and the fact that the phone was discontinued on May 15, we would think you are due us one-half month return. In other words, a credit of $17.75. We wish to make some reasonable agreement with you for the telephone depot at Chireno. We would think that $5.00 per month would be satisfactory, and ask that you take into consideration that eight miles of the line between Chireno and Lufkin was put in by us and is at present our property."

On August 29, the lumber company wrote Judge Mantooth, apparently in reply to his letter of August 28: "I have carefully read your letter...with reference to the charge made against our company for services on the Chireno line. For your information, we do not use the line to our depot at Chireno over three times a day. This I would say is the maximum from the railroad office at Keltys to the depot at Chireno. Further call your attention, that out of 30 miles of line, we built something like eight miles ourselves, and your exchange is using this line right along to get outside tolls which should be considerable. You seem to be under the impression that you do not get outside toll calls on the Chireno line."

Southern Pine Lumber Company workers gather in front of the old Diboll Commissary store in the 1920's. Photo, T. L. L. Temple Library.

In 1924, with the new Stromberg-Carlson equipment in use, Judge Mantooth, in response to a letter from Southwestern Bell, summarized the Lufkin exchange's service area, including a 10-mile toll line from Lufkin to Huntington, connecting with the Huntington Exchange and then to Broaddus, with a branch from Broaddus to Camp Bannister, about three miles long, extending from Broaddus to White City. Mantooth said the company had abandoned a line from Lufkin to Ratcliff and then to Alcedo along the old Eastern Texas Railroad. He said the exchange had a toll line from Lufkin down the HE&WT railroad connecting with Burke and Diboll, then along the route of the Texas Southeastern Railroad to Vair, then along the tramroad of the Southern Pine Lumber Company to Alcedo. Another extended from Lufkin on the right-of-way of the Angelina and Neches River Railroad to Chireno in Nacogdoches, a distance of 30 miles.

"This constitutes the toll lines, and the only toll lines, of the Lufkin Telephone Exchange," said Mantooth. "We have some rural lines extending out from Lufkin in different directions of the county, but none in competition with any of your lines."

On February 24, E.L. Kurth, president of the Angelina County Lumber Company, complained to Mantooth that his company had been overcharged for telephone service between Lufkin and the controversial logging camp in San Augustine County. "We do not pass over three calls a day to Chireno now, and certainly you can not expect us to pay $25.00 per month for this small service. We will be glad to pay $10.00 per month in order to settle the controversy and will agree for our section men to keep the line up from the Angelina River to Chireno, or we will pay you at the rate of 15 cents per call, as you seem to think that $22.00 per month is a good deal less if you were to charge us at the rate of 15 cents per call, or if this is not reasonable, we will be willing to purchase at some reasonable price the portion of the telephone line that we do not own, and which is owned by you, for as you know, something over eight miles of the line was constructed and paid for by ourselves."

Mantooth, in a reply to Kurth on March 4, agreed to charge $10 per month with Kurth's men maintaining the line from the Angelina River to Chireno. That apparently ended the logging camp phone controversy.

In February of 1925, Judge Mantooth received a proposition from Frank Scope of Henderson Telephone Company to join in the consolidation of several East Texas exchanges. Scope proposed that an operating company lease the local facilities and allow each individual telephone plant to bear the burden of its maintenance. Mantooth replied he would be willing to consider a consolidation, but added: "I am of the opinion that I have the best telephone exchange in Texas." The consolidation issue was never pursued by Lufkin.

On March 20, the Lufkin exchange filed a lawsuit against the City of Lufkin, attempting to garnishee the wages of a city employee for failing to pay a $7.95 telephone bill.

On February 20, 1926, the telephone company reported capital stock amounting to $75,000, bills payable of $5,500, phone rentals of $42,231, and tolls and sundries of $29,052. The company reported a profit of $2,059 for the month.

In August of 1927, J.O. Satterwhite, a friend of Mantooth, raised the issue of purchasing and sale of independent systems contiguous to Lufkin. He wrote: "Rusk and Dialville have signed up in keeping with the telegrams shown to you, and Mr. Cammack's engineer and inspector has put his OK on Alto and Groveton...Mr. Cammack has said he believed his brother would agree to a figure as high as $235,000 for the Lufkin system..."

In the 1928 edition of Polk's City Directory for Lufkin, the telephone exchange was located at 23 1/2 Dozier Avenue (the second floor of the Denman Building) with E.J. Mantooth as president, Mrs. A.M. Denman as vice-president, and Jean Shotwell as secretary-treasurer and manager. The company's telephone number was 823. Southwestern Bell was also listed as an occupant of 23 1/2 Dozier Avenue. The directory further reported that E.L. Kurth was the president of the Lufkin Chamber of Commerce.

In 1929, the company's board was broadened with Judge Mantooth as president, Kester W. Denman,Sr., was secretary-treasurer, and N.D. Shands, L.H. Denman, D.C. Hucherson, and Mrs. A.M. Denman as directors. Jean Shotwell remained as general manager. The board remained unchanged until 1939.

Lufkinites watch a parade in front of City Hall when it was located on Cotton Square. Photo, Bob Bowman Collection.

Conroe's telephone exchange was located in this second-floor room in the Madeley Building in downtown Conroe.

8 Creating the Lufkin-Conroe partnership

The towns of Conroe and Lufkin, both located in the East Texas forests, have much in common.

Both trace their beginnings to the railroads which began to crisscross East Texas in the early 1880s, linking the region to the shipping ports along the Gulf Coast. Both received their early economic impetus from lumbering. And both owe their evolution to pioneer families which settled in their counties after Texas won its independence from Mexico.

A year after Dr. Alexander Madison Denman and Judge Edwin James Mantooth—the products of pioneer families in Angelina County—installed the first telephones in Lufkin, two other early businessmen did the same in Conroe.

Conroe had been served by a long-distance line from Houston to Gilbert's Drug Store for several years and in 1899, Albert A. Madeley and a man named Gilbert conceived the idea of creating local telephone service. By 1900, a switchboard had been installed in the rear of Gilbert's Drug Store with P.L. Curd and S.K. Hailey as the operators.

Shortly thereafter, Albert's brother, George B. Madeley, purchased the telephone company from the two previous owners, named it the Conroe Communication Company, and moved it to the second floor room in the Madeley building at 125 Simonton Street, just off the Montgomery County courthouse square in downtown Conroe. The lower floor was occupied by a meat market operated by his brothers, Albert A. and Daniel A. Madeley, who had large herds of cows roaming over a large portion of Montgomery County, which was open range at the time.

George Bell Madeley, the patriarch of the Madeley family in Montgomery county. Photo courtesy Montgomery County Heritage Center.

The patriarch of the Madeley family in Montgomery County was George Bell Madeley. He was born August 13, 1815, at Trittiford in Hall Green, a suburb of Birmingham, England, but immigrated to the Republic of Texas to seek his fortune. In Texas, he married Helen Grant, the daughter of a fellow Englishman, in 1847. They purchased 320 acres of land at Egypt in Montgomery County, established their home and reared seven sons in an area that was virtually a wilderness. Conroe was nonexistent and the closest settlement was Houston, 35 miles to the south.

Madeley was an enthusiastic and innovative farmer. He grew an orchard, an eight-acre vineyard stocked with vines he brought from England, a sugar mill, a wine press, cotton gin, herds of cattle, and a gristmill.

Early in his farming career, the Montgomery County area sustained a severe drought and Madeley's neighbors did not have seed corn because of their crop failures. They came to Madeley to purchase seed and named his farm "Egypt" because of the Biblical story in which Jacob's family traveled to Egypt to buy corn from their brother Joseph in a time of famine.

Houston lumberman Isaac Conroe, who gave Conroe its name.Photo courtesy Montgomery County Heritage Center.

Madeley was described as a "sugar planter" in the 1860 Montgomery County census. He was known widely for his wines, which he shipped in wooden casks to his family in England. The wine may have been the first in Texas to be exported.

During the Civil War, when the Port of Galveston was blockaded, Madeley formed an oxen train and hauled his cotton and that of his neighbors to Brownsville, a trip requiring two months. He also ground corn and ginned cotton for neighbors who were away at war.

Madeley died March 22, 1879, and was buried in the family cemetery at Egypt. His wife Helen was also buried there in 1897.

In 1899, when Albert A. Madeley and Mr. Gilbert created the town's telephone service, Conroe was Montgomery County's largest community, but it had less than 700 residents. Most of the town's businesses were clustered in a square around the county courthouse, which had replaced Montgomery as the county seat in 1889.

Conroe owes its name to Houston lumberman Isaac Conroe, who owned a small sawmill at Haltom south of Conroe, and moved the plant to the J.G. Smith survey, not far from the International and Great Northern Railroad's Houston-to-Crockett line. He built a two and a half-mile railroad tram from the mill to the I&GN Railroad and used mules to pull the tram cars hauling lumber from his mill to the main line.

A post office was established in the sawmill town in 1884 and took the name of Conroe Switch in honor of the Northern-born former Union Calvary officer who served as its first postmaster. Within a decade, the name was shortened to Conroe.

When Conroe became the county seat in 1889, Isaac Conroe allowed the county to use the downstairs of his home at the corner of First and A Streets to serve as a temporary courthouse until a permanent structure was built in 1891.

The prosperity of the local timber and agriculture industries in the early 20th century enabled Conroe to grow quickly. But, like turn-of-the century Lufkin, two fires destroyed portions of the town. One fire occurred in 1901 and the other flared up in 1911.

The Madeley Building housed Conroe's early telephone exchange on its second floor. The name on the building is misspelled.

After years of sustained growth, the town's prosperity was threatened in the late 1920s when a dwindling timber supply closed down many of Montgomery County's sawmills.

In 1928, George B. Madeley agreed to sell his 29-year-old telephone company to the Denman and Mantooth families of Lufkin. The company was still housed in the Madeley building at 1225 Simonton Street, but the meat market on the first floor was later replaced by a dress shop owned by Mrs. W.M. Bell and her son.

Judge E.J. Mantooth negotiated the acqusition of the Conroe telephone exchange from the Madeley family. He gave his grandson Jean Shotwell, who became manager of the Lufkin exchange, an interest in the exchange worth $1,000.

The sale was culminated in early 1929. In a letter dated January 20, 1929, H.B. Finch of the Union National Bank of Houston wrote Judge E.J. Mantooth of the Lufkin exchange: "This will acknowledge receipt of escrow agreement executed by G.B. Madeley, Mrs. G.B. Madeley, Neil Madeley, E.J. Mantooth, and N.D. Shands, bearing date of January 9, 1929, together with a bill of sale and two drafts of $20,000 and $16,749, respectively; also stock bonds containing certificates 1 to 100 inclusive of the Conroe Communication Company..."

The Conroe exchange was reorganized as the Conroe Telephone Company with the board consisting of E.J. Mantooth as president, N.D. Shands as first vice-president, L.H. Denman as second vice-president, Kester W. Denman, Sr., as secretary-treasurer, and D.C. Hutcherson and Mrs. A.M. Denman as additional directors. Jean Shotwell was named general manager and assistant secretary-treasurer. The board remained intact until 1939.

At a called meeting of the Conroe City Council on January 12, 1929, the Conroe company, now in the ownership of the Denman and Mantooth families, was granted a "renewal and extension" of a franchise that had been granted on September 11, 1907. The company was also given a new 50-year franchise dating from September 11, 1932. On March 21, 1929, Judge Mantooth received a new charter for the Conroe company from Jane Y. McCallum, the Texas Secretary of State, in Austin.

The 1929 Conroe system consisted of a Kellogg six-position switchboard, which had two toll operators, one rural operator, and three local operators.

The new owners planned to make improvements to the Conroe system, but the Great Depression put an end to their plans. In 1930, Conroe's only bank failed and pushed many residents and businesses into financial doldrums, the town's lumber industry remained depressed, Conroe's schools struggled to complete full terms, and hundreds of people were without jobs.

The town's fortunes began to improve some in 1931 when George W. Strake discovered oil seven miles southeast of town, triggering an oil boom in the county. Within weeks, the town's economy was revived and at one time, Conroe reportedly claimed more millionaires per capita than any other town in the U.S.

A switchboard of the type used by early telephone operators in Conroe.

Mantooth (Bug) Shotwell, Jean Shotwell's brother, who managed the exchange for five years in the 1940s, remembers the discovery. "There wasn't much to Conroe until they found oil. In those days, you had to wear boots because the downtown streets weren't paved, and there was mud everywhere," he said.

Mantooth remembers staying at the old Charles Hotel while working in Conroe. "The room and board there was $1 a day, and for another dollar you got the best meals in town," he said.

Even though Conroe's economy was improving, the Lufkin Telephone Exchange remained under financial restraints and capital for expansion in Lufkin or Conroe was limited. The Lufkin economy—where much of the owners' resources lay—was still in the grips of the depression. Many of Angelina County's sawmills had closed, the businesses that depended on lumbering were floundering, and jobs were scarce.

At the start of 1934, Conroe had only 372 telephones—including homes, businesses and extensions—and the number fell to 364 by October. At Achisonville, the exchange's only outlying community, the number fell from 404 to 384. In Lufkin, the year began with 1,480 telephones and had risen to 1,537 by November.

Some of the early managers of the Conroe exchange were Minnie Payne, Danny Yarbra, and Mantooth (Bug) Shotwell.

An early pay telephone used in the Conroe exchange. The phone was answered by an operator.

The forties were difficult years for the Conroe exchange. In 1941, the Conroe City Council approved a new ordinance establishing rental fees to be paid for telephone, telegraph, electric, and gas companies for the privilege of using the city's streets, easements, alleys and other public ways in the city. The companies were mandated to file with the city tax department a sworn report showing their gross receipts exclusive of the sale of merchandise.

In 1947, more than 90 Conroe citizens submitted a petition to the Conroe City Council, protesting the exchange's service. The group contended, "As a part of the consideration for the granting of the franchise...the Conroe Telephone Company agreed to enlarge and improve the telephone system and to furnish adequate telephone service for the citizens of Conroe." The petitioners said that during World War II, the Conroe exchange "offered as an excuse for its failure to render adequate and satisfactory service, the fact that there was then a shortage of material and equipment with which to repair the telephone system and to enlarge and improve the system."

In August of 1947, Jean Shotwell reported to his board of directors at Lufkin on the problems in Conroe. "As you know, at the time of the purchase...of the Conroe Telephone Company...the people of Conroe were not telephone conscious," he wrote. "The (telephone) plant was small, in bad state of repair and had far too few customers for a town the size of Conroe."

Shotwell reported that with the discovery of oil in Montgomery County, "Conroe began to grow from a business and industrial standpoint, as well as population...(and) along about 1939, we saw that the town had outgrown and was continuing to outgrow our plant and its facilities there, and that it would be necessary to expend large sums of money in order to have the plant in such condition and of such size to meet the needs and requirements of a growing community."

Shotwell continued: "For the past several years, we have been watching the growth of Conroe, and it soon became evident that the town was growing...in a southwesterly direction from the courthouse...and that our main leads and most of our business would be in that area."

Shotwell said the Conroe facilities were "too small and cramped, inconveniently located and unsuitable in every respect." He added that the company had acquired a lot measuring 150 by 100 feet on Phillips Street in downtown Conroe, where it planned to create a new telephone facility for Conroe.

"During the war, beginning in January, 1942, the Conroe Telephone Company purchased $19,500 (maturity value) of war bonds. Up to this time, $6,000 of these bonds have been sold and an additional $6,000 is in the process of being sold.

"In line with our plans made several years ago for recabling the City of Conroe...all poles, messengers, and hardware necessary, not only along the lines of the old plant, but new construction in new additions have been paid for during this past 17 months. Shotwell said the cable for the improvements were to be delivered within the next 12 months. "The cable capacity of the plant in Conroe...will be doubled and will provide the city...with adequate telephone service," he told the Lufkin board.

Morris Bateman, who became manager of the Conroe telephone exchange in the 1940s, works on a test board in 1949.

In 1948, following the resignation of Mantooth (Bug) Shotwell as manager of the Conroe exchange, Jean Shotwell turned to Morris Bateman, the wire chief of the Lufkin exchange and the husband of his secretary, Marjorie, to manage the Conroe operations. Bateman, who had joined the company in 1938 as a lineman and installer, arrived in Conroe on August 1, 1948. At the time, the exchange had a six-operator Kellogg switchboard to serve the entire exchange, Mary McNaulty was the chief operator, and the company was still renting office space from the Madeleys on Simonton Street. "I carried the rent check to Mrs. Madeley (the widow of the exchange's founder) every month," recalled Bateman.

"There wasn't a house anywhere in town to rent, so I had to rent a room at the old State Hotel while Marjorie stayed in Lufkin," said Bateman.

Later, Bateman moved into a "great big old country home" that stood on a lot at 300 Phillips Street, where the telephone company had purchased the downtown lot for building a new telephone building. The house had been vacant. The company "fixed up the house," said Bateman, and he lived there for six years until he could build his own home.

"They named me manager of the exchange, but being manager meant doing everything that needed to be done," said Bateman. "It also meant giving the people somebody to fuss at."

Bateman had been in Conroe less than a year when another flareup of customer discontent occurred. According to a story in the *Conroe Courier* of August 11, 1949, the Conroe City Council, in response to additional complaints from telephone customers, "passed a motion...directing a study of the Conroe Telephone Co. with a view to bringing action to cancel the company's franchise in Conroe."

The story added: "The motion was made by W.A. (Bay) Evans following discussion of the increasing number of complaints from telephone subscribers. The complaints not only covered service, but also were reported by Mayor Earle Gentry to include what would be considered unusual in determining what rate should apply...the motion is the culmination of complaints and dissatisfaction that have been before the council the past two years. Similar motions were considered at other periods, and the company officials made promises of better service and installation of telephones. The installations were made after months of delay due to a lack of cable, but that is as far as the company went, according to the complaints."

Two days after the *Conroe Courier* story, and a similar story in *The Houston Post* and *The Houston Chronicle*, Dr. Byford H. Denman of Lufkin, a stockholder in the Lufkin and Conroe telephone companies, wrote Mayor Gentry, pledging improvements in the Conroe system. "I see no reason why our company cannot continue in your fast growing city and give you proper service you want and are desiring of," said Dr. Denman, a son of company founder Alexander Madison Denman.

The Conroe Hotel, an early turn-of-the century landmark in the community. Photo courtesy Montgomery County Heritage Center.

Conroe City Attorney George B. Darden replied to Dr. Denman's letter: "All the city desires is a modern, up-to-date telephone system with less delays as possible on long distance calls, and we hope that this can come about and that the service generally will be improved...your Mr. Jean Shotwell has recently been in our midst and assures us that this will be done."

When Bateman arrived in Conroe, the town was still largely a lumbering and ranching town. "It was a small country town with one bank and several grocery stores on the square. Saturday was a big day, everybody came to town, and did all their shopping on the square. The doctors' offices were upstairs in the buildings on the square."

To increase telephone service in the community, Bateman expanded the switchboard and added two operators, one for local service and one for long distance. Because there wasn't sufficient room in the cramped upstairs office to put the board against the wall, he lined it up from corner to corner.

Doris Johns of Lufkin, who worked in Conroe between 1942 and 1986, remembers the old office well: "The office was cramped, dark, and there was a dress shop below."

One of Bateman's biggest problems was providing long distance service. "We had only six circuits to Houston and at 8 a.m., people would start calling in with their requests for long distance to Houston. The operators started placing calls and by 10 we'd have a stack of tickets a half-inch high from people wanting long distance. To get long distance, you had to go to Houston, whether it was Chicago or

A group of men gather on the porch of a Conroe drug store around the turn-of-the century. Photo courtesy Montgomery County Heritage Center.

anywhere else."

Dave Nichols, who owned an automobile supply store in Conroe, once called the exchange for long distance service, hoping to order repair parts from a Houston firm. The operator put his request on the pile and called him back several hours later to advise him that his long-distance call was ready. "I've already been to Houston and picked up the dang parts," he told her.

All of the exchange's long distance calls went over Southwestern Bell lines, which ran along the railroad tracks south to Houston. "When the lines were along the tracks, I never got more than 12 circuits to Houston," recalled Bateman. "Later, Bell put a cable along the Houston highway, and our long-distance service got better."

Bateman's efforts to solve the Conroe exchange's problems were hampered by insufficient equipment. In a town of 5,000 people and 800 customers, the exchange could handle only 80 conversations on the city switchboard and 20 on the rural board at one time. "By the fifties, we were in deep trouble," recalled Bateman, "we had more telephones than places on the switchboard, which the operators needed to connect the calls, and we didn't have the money to make the kind of improvements that were needed."

Doris Johns recalls those difficult days: "A lot of people from Houston were moving to Conroe because it was a nice little country town, but they expected telephone service like they had enjoyed in Houston. We were operating in a smaller town, and we just couldn't do what they wanted in the way of phone service."

At one point in negotiating with the Conroe City Council, Mayor Bill Newton, a wealthy oilman, told Bateman and Marion Clay, who had succeeded Jean Shotwell as manager of the company, that he was prepared to buy the Conroe exchange himself. "The mayor slapped his checkbook on the table, and told us to give him a price, but we didn't have the authority to do so," remembered Bateman.

"It was right after the war, and everybody wanted a phone, so I changed out the switchboard so we could have more party lines. We had two-party and four-party lines in town and eight-party lines in the rural areas," he said. "Around town, they got to calling me Party Line Bateman."

Service in the community continued to be so tenuous that Bateman had to frequently monitor the operators' work from his home, using a direct line to the switchboard. "I'd sit at home and listen in on the conversations to make sure our customers were being treated right," he recalled.

On one occasion, Bateman was at home when a customer came to his house late at night, complaining that the operator didn't answer the telephone. Bateman listened in on his private line, heard nothing, and drove down to the company offices. Entering the switchboard room, he found the operator sleeping on several chairs she had pushed together to form a makeshift bed. He roused her from her sleep.

Doris Johns said Bateman worked hard to convince Conroe subscribers he was trying to do everything he could to provide them with quality telephone service. "There wasn't anything Morris wouldn't do. If someone asked him to change a light bulb, he'd do it. He never acted like a big executive, and he worked real hard doing anything that needed to be done," she recalls.

Angelina County residents dedicated in 1936 a state historical marker at the ghost town of Homer, where the Denmans and Mantooths were prominent in business. The dedication was made during the Texas Centennial.

9 A depression and the war years

The Great Depression of the 1930s spread across East Texas like an early-morning fog. Many of East Texas' principal industries closed their doors or curtailed operations, dozens of banks collapsed, and the value of real estate and other investments plummeted. Ripples spread through the economy, affecting almost everyone.

Lucy Hendrick of Lufkin, who worked 36 years with the company, remembers the late thirties as "hard times." But she recalls that "everybody had the same problems, and we all managed to get along."

While the Lufkin and Conroe telephone systems continued to show modest growth in the number of homes and businesses they served, funds for capital improvements were scarce and many of the improvements the Denman and Mantooth families felt were essential to building customer bases in Lufkin and Conroe had to be postponed.

The Conroe system was in particular need. Customers complained frequently about the telephone service. In a 1932 petition submitted to the Conroe City Council, customers said there "is no privacy of telephone conversations" and complained that "any subscriber who has a private line can at any time pick up his telephone instrument and listen to one and sometimes as many as four conversations going on between persons on other lines in the city."

"This deplorable situation has led many businessmen in the City of Conroe to conclude that business matters should not be discussed over the telephone," the petitioners claimed.

An early photograph of the switchboard which served Lufkin before dial service came to the community.

While the Lufkin and Conroe systems managed to stay in business during the worst years of the Depression, it was a struggle to keep the systems' employees on the payroll. Profits were meager and funds for improvements were almost nonexistent.

A balance sheet for the Lufkin system for October, 1934, showed a monthly net profit of less than $995 and a net growth of only 57 since January. The balance sheet for the same month in Conroe showed a net profit of $318 and a loss of 20 customers since January.

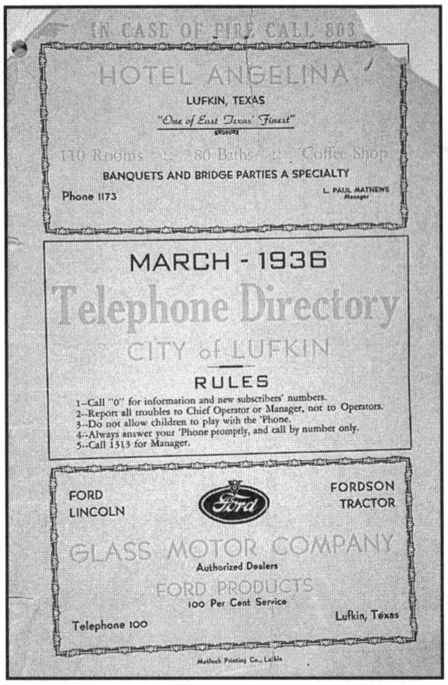

Lufkin's telephone directory in 1936.

Even during the tough times of the 1930s, the telephone systems remained essentially a family business for the Denmans and Mantooths, and their attitude extended to the men and women who worked for the company. The company's concern for its employees is reflected in an August, 1934, decision to create a retirement plan for the Lufkin and Conroe employees. It was a remarkable and compassionate decision, considering the economic uncertainties of the 1930s. In a letter to all employees on August 1, 1934, general manager Jean Shotwell wrote: "The problem of old-age dependency is a serious one and the company feels that its employees should make provision for an income during their old age."

In 1935, the Lufkin exchange had 23 employees serving 1,567 stations, and handled some 13,000 calls a day. The same year, Conroe had 400 telephones in service.

By 1938, the Depression had lessened and the Lufkin and Conroe systems were showing a financial improvement. On October 30, the Lufkin exchange reported a net profit of $2,731 for the month and a growth of 18 customers since January. Conroe reported a net profit of $708 and an increase of 33 customers since January. The Conroe growth in customers can be attributed largely to the discovery of oil in Montgomery County, which spurred Conroe's growth in the late 1930s.

In 1939, the board of directors of the Lufkin exchange approved the creation of "a reserve fund or surplus to be used in making many repairs" to the Lufkin system, to meet the increasing growth of Lufkin, and to safeguard the system against risks "such as cyclones, ice storms, fires and other like casualties," and to "prevent having to borrow money...which would have to be repaid and which the Board deemed excessive under the circumstances." The board set aside five percent of its net earnings for the fund.

In February, 1939, the board of the Lufkin exchange added H.T. Bowers and Shirley Simons as directors, succeeding Mrs. Alexander Madison Denman, widow of the company's co-founder. In December, Peyton R. Denman succeeded D.C. Hucherson as a director. In Conroe, the board was made up of E.J. Mantooth, president; Kester W. Denman, vice-president; N.D. Shands, Jr., L.H. Denman, Simons, Bowers, and Peyton R. Denman, directors. Jean Shotwell remained as manager and secretary-treasurer. The board remained unchanged until 1941,

Telephones such as this were often found in Lufkin and Conroe homes in the 1930s and 1940s before dial service came to the two communities.

when D.C. Hucherson was added. In 1941, the Lufkin board consisted of Mantooth, Shands, L.H. Denman, Hucherson, Peyton R. Denman, Bowers, Simons, and Kester W. Denman, Sr.

The late 1930s was the start of an extraordinary era of growth in Lufkin's history. Lufkin Foundry and Machine Company was expanding to keep pace with a growing market for its oilfield pumping units. Texas Foundries, Inc., the first foundry to be located west of St. Louis, was located on North Raquet Street and began producing foundry castings in 1939. And, across town in the emerging settlement of Herty, Southland Paper Mills, Inc., the first paper mill to produce newsprint from southern pine wood fiber, was nearing completion. The mill began making newsprint in January of 1940.

The Denman Building as it appeared in the 1990s. The building was built around the turn of the century and remodeled in 1939.

Even as the telephone exchange grew in Lufkin, the company remained in its second-floor offices in the Denman Building, owned by Dr. Denman's widow. In 1939, however, the building underwent a remodeling that replaced the wooden exterior with a brick one and enlarged the telephone company offices.

"All of the telephone offices were in this small area above a drug store, and we needed more room. We had only two offices, a little storeroom, a battery room and a switchboard area," recalled Morris Bateman, who worked in Lufkin until 1948 when he was named manager of the Conroe exchange.

A common battery telephone used in the 1930s. The telephone was answered by an operator.

"While they were remodeling, we had to keep the switchboard, the mainframe, and the operators working, so they put some steel beams underneath the switchboard and built a complete box around the switchboard with tarpaper walls and a tarpaper roof. The operators would climb a set of dinky stairs, and go inside this box to work their shifts," said Bateman.

Working conditions in the box were so hot, Bateman said, that tubs of ice were put inside the room with the operators and fans were set up to blow across the ice to keep the operators and the equipment cool.

One of Bateman's jobs during the remodeling was to clean the construction dust off the switchboard each night so it would operate properly the next day.

"They built a new building, the one standing there today, right around that box. And when they were finished, they eased the telephone equipment down on a new concrete floor and tore out the tarpaper box," he said.

In 1942, a telephone repairman accidentally started a fire on a telephone pole just outside the new second-floor telephone office and knocked out every telephone in town. The fire burned five major cables feeding into the switchboard. "Every line coming into the office was burned, and we called Nacogdoches and got some help, and six of us spliced wires for two days and two nights to get the system back on the line," said Bateman.

Even though the Denman building had been remodeled, Betty Warren, who worked there in the 1940s, remembers that "we could look through the slats in the floor and see what was happening in the Ruby Cafe downstairs."

Lufkin telephone operators take a lunch break at the old Cash Drug Store in the 1940s. Left to right are Doris Redd, Lucy Hendrick, Evelyn Free, Billie Nerren, and Maxine Parrish.

In those days, Mrs. Warren recalls, everyone in town paid their telephone bill in person and had to climb a set of stairs to the second-floor telephone office. "The stairs went straight up and were very difficult to climb," she said.

Lufkin's economy boomed and the town's telephone service was expanded in the late 1930s and early 1940s, but in 1942 the United States entered World War II. At that time, it became evident to the telephone exchange that the war would prevent the purchase of materials for the expansion and repair program embarked on in 1939. As a result, the board increased its reserve fund to meet the needs that would be created once the war ended. One step the exchange took was the purchase of $38,000 in war bonds, of which $15,000 were earmarked for future expansion and repairs.

In 1942, the Lufkin board was made up of Judge Edwin James Mantooth, who was president, N.D. Shands, L.H. Denman, Dave Hucherson, Peyton R. Denman, H.T. Bowers, Shirley Simons, and Kester W. Denman, Sr., who was vice-president. On the Conroe board were Mantooth, president; Shands, L.H. Denman, Hucherson, K.W. Denman, Sr., vice-president; Simons, Bowers, and Peyton R. Denman. Jean Shotwell remained as manager of both exchanges, secretary-treasurer of Conroe and assistant secretary-treasurer of Lufkin.

When the war ended, Lufkin began to grow again, almost doubling its population between 1939 and 1947. The exchange also had to move its lines in Diboll from the Texas-New Orleans Railroad line because of a federal court decision and was forced to rebuild almost all of its facilities north and northwest of Lufkin, the result of an ice storm in 1943.

Just as it had been in the 1920s and early 1930s, the Lufkin exchange remained a strong supporter of Lufkin. In 1938 and again 1939, company floats that were entered in the Texas Forest Festival Parade, the community's biggest annual event, won first prize in the Festival parade. Jean Shotwell, the company's general manager, was president of the Festival in 1940.

In 1941, the company found itself involved in a dispute with a Lufkin chiropractor. In a resolution passed by the company, it was alleged that the doctor, his wife, who owned a beauty parlor, and their daughter, a public stenographer, had been using a residential telephone for business reasons. "Said telephone has been discontinued and disconnected for the reason that Dr. ...was not agreeable to paying a business rate or charge," said the resolution. The controversy was resolved when the daughter was given a residential phone after she promised it would not be used for business by the family.

In 1944, according to company retiree Jean Abbott, the cost of basic local service in Lufkin was $2.50 with a 38-cent tax added. The business rate was $5.75 with an 86-cent tax. Rural party service was $1 with a 15-cent tax. The Lufkin exchange's board that year was made up of Mantooth, president, Kester W. Denman, Sr., vice-president; and directors L.H. Denman, D.C. Hucherson, Shirley Simons, H.T. Bowers, Peyton R. Denman, and Mrs. N.D. Shands Sr. The Conroe board had the same directors.

In 1944, employees at Lufkin were saddened to learn of the death of Judge Mantooth, who had founded the company in 1898 with Dr. Alexander Madison Denman. The judge had been in declining health for many years and from their window at the corner of North First and Dozier, they often saw the judge step out of his chauffeur-driven Packard and enter the Lufkin National Bank building, a block away, where he maintained an office on the bank's second floor.

With the death of Judge Edwin James Mantooth in 1944, Kester W. Denman, the son of telephone company co-founder Alexander Madison Denman, became president of the Lufkin exchange, as well as its sister exchange in Conroe.

The judge, who had been the telephone company's acknowledged leader since the turn of the century, was laid to rest in Hillcrest Cemetery in one of the town's biggest funerals. His civic and business contributions to Lufkin and Angelina County were recognized by everyone.

With Judge Mantooth's death, Kester W. Denman, his law partner, became president of both the Lufkin and Conroe exchanges and Mrs.Dell Lanter was added to each board.

In the 1940's, employees at Lufkin and Conroe worked long hours to keep up with the growing number of customers. T.D. Yarborough of the Conroe exchange wrote the following letter to the Lufkin exchange in 1945 to explain his work hours.

"Your letter of July 31 is before me. I can't quite understand how to adjust a fact. Mr. (Jean) Shotwell was over on the 2nd, but was busy and I did not feel like bothering him; hence, this letter...I can't see how you good people would doubt that I worked on the 27th, especially since you all were here that day. On this particular day (the 27th), I came to work at 2 a.m. It must be realized that there is a central office to be kept up and especially if there is trouble, it must be remedied before the girls (the operators) come on at 7 a.m. I still contend that it is far better to take care of the inside in the early hours and devote the day to trouble and work that must be done. On Friday, I had an unusual amount of small cases of trouble, some of which involved cable trouble. After I received my call from Kilgore, I worked all the harder to get the trouble up before I left. Mr. Smith and I worked through dinner to hurry and get through. At 2:30, I took Henry home to dinner and came to this office and reported to Miss Ola that I was fixing to pull out. I went to the storeroom and repaired a phone for Mr. Smith...leaving the storeroom at 3 p.m. I still contend that from 1 a.m. to 3 p.m. is 13 hours. As for the 28th, I did not turn in any time for I was in Kilgore until 8 a.m. The doctor says that my son was definitely on the road to recovery as the penicillin had finally began to be felt. He said that he wanted him to get all the sleep he could Saturday night and Sunday, and I knew if I stayed in the room, neither he or nor my wife would sleep. I left Kilgore at 8 p.m. and drove home, arriving here at 12:30 p.m. I immediately reported to the operator that I was back. On Sunday, I came to work at 6 a.m. and was in and out of this office practically the entire day up until about 3:30 P.M., at which time I went home for 12 hours of much needed rest. I am sorry, but I cannot see where I can adjust these days for they speak for themselves."

Dr. L.H.Denman, who served as president of the Lufkin telephone exchange in the 1940s, 1950s and 1960s.

The year 1945 also saw Kester W. Denman, Sr., reelected as president of both the Lufkin and Conroe exchanges. Serving on the boards were D.C. Hucherson, vice-president; L.H. Denman, Shirley Simons, H.T. Bowers, P.R. Denman, Mrs. Dell Lanter and Mrs. N.D. Shands Sr. Jean Shotwell remained as manager and secretary-treasurer of the two exchanges.

The boards remained unchanged until 1948 when L.H. Denman was named president, N.D. Shands Jr. was chosen first vice-president, and H.T. Bowers was elected second vice-president. The other board members were D.C. Hucherson, Shirley Simons, and Peyton R. Denman.

A 1946 report on wages paid to employees of the Lufkin and Conroe exchanges showed that a typical employee earned 48 cents an hour, worked 20 to 24 days a month, and earned about $100 a month.

Lufkin's industrial growth in the mid-forties also put a strain on the company's ability to serve new customers. On March 21, 1946, W.W. Trout, vice-president and general manager of Lufkin Foundry and Machine Company, wrote Jean Shotwell, asking for a telephone for a new employee, H.A. McKerral. "I realize there are more people looking for telephones than you can keep up with...I believe Lufkin Foundry for years has been the largest customer in your system and I believe we are entitled to a little bit more consideration than to have to wait four or five months for a phone when plenty of people are getting them."

Shotwell wrote Trout: "Walter, you don't seem to remember for the past three years how many of your key men we have furnished with telephones and the two...systems we installed for the foundry, which I had to practically beg out of the manufacturer. In fact, the foundry or foundry employees have been furnished with about 20% of the telephones we were able to get in the past three years." Shotwell concluded his letter by promising Trout that "Mr. McKerral will get a telephone as soon as we have a line close to 349 Everett Street (McKerral's home)."

A 1946 letter to Kester W. Denman, Sr., from another Lufkin industrialist, Cal C. Chambers, president and general manager of Texas Foundries, Inc., provided an insight to Lufkin's economic successes in the forties: "The end of the coal strike brought assurances that Texas Foundries will continue to operate at full capacity the remainder of the year, except during Christmas week when we will close down to

Dr. L.H. Denman's landmark home on North First Street in Lufkin. The home was designed by well-known architect Shirley Simons of Tyler.

install a new cupola and make some routine year-end mechanical repairs. The year 1946 will be our biggest and best year ever since this company was organized and we go into 1947 with a very substantial backlog of orders and high hopes for a repetition of this operating and financial success..."

A similar letter in 1947 to Denman from R.W. Wortham, Jr., executive vice-president of Southland Paper Mills, Inc., offered additional economic encouragement. "During 1948...we will have our second paper machine operating and this will provide us with several thousand more tons of production annually...at the same time, we do not anticipate any reduction in market prices on newsprint. In view of this, our earnings next year should be considerably greater than at present," said Wortham.

Buoyed by Lufkin's strong economy, Dr. Peyton R. Denman of Houston wrote his brother Kester W., both substantial shareholders in the Lufkin exchange, with a suggestion in 1946 that the company's board increase its rates and raise its capital stock to $150,000. The company's shareholders in 1946, in addition to Dr. Denman and Kester W. Denman, Sr., were Mrs. Nell Denman Abram, H.T. Bowers, Mrs. Robbie Bowers, B.H. Denman, Mrs. Cleo Denman, L.H. Denman, Mrs. Jeannette M. Harrison, Mrs. Archie Harrison, D.C. Hucherson, Mrs. Dell Lanter, Mrs. Merle Starr McMullen, Mrs. Muriel Joan Mantooth, Mrs. Ned D. Shands, N.D. Shands, Jr., Jean Shotwell, Mrs. Winnie Shotwell, Mrs. Mollie Simons, Shirley Simons and Mrs. D.C. Hucherson.

In 1947, general manager Jean Shotwell reported to the exchange's board of directors: "The rapid growth of Lufkin and its environs since January 1, 1942, has greatly overtaxed the plant. We have hundreds of applications on file for telephones and are unable to render the service. Upon completion of new additions to the switchboard and line facilities, relief from this congested condition will be obtained, temporarily, at least. If the two-party system which we plan to establish is not met with too much opposition, our plant facilities will be adequate for a number of years."

In February of the same year, the Lufkin City Council followed the example set by the Conroe City Council in 1941 and established a 1.5% tax on the Lufkin exchange's gross receipts. Like the Conroe ordinance, the Lufkin action covered the use of streets for poles, conduits, cable and other telephone facilities. City officials estimated the tax would produce an income of $1,750 for the city each year.

In March of 1947, the Lufkin council also approved, at Shotwell's suggestion, an ordinance establishing local telephone rates. Shotwell explained that telephone service was the only public utility in Lufkin not covered by a city ordinance setting maximum rates. The Council's March ordinance set a maximum rate of $5.50 per month for a one-party business rate, $5.00 for a two-party phone, $3.00 for a one-party residential phone, $2.50 for a two-party phone, $1.25 for each business telephone extension, $1.00 for each residential extension, $8.25 for each trunk line, $3.50 for each business straight line installation, and $2.50 for each straight line residential installation.

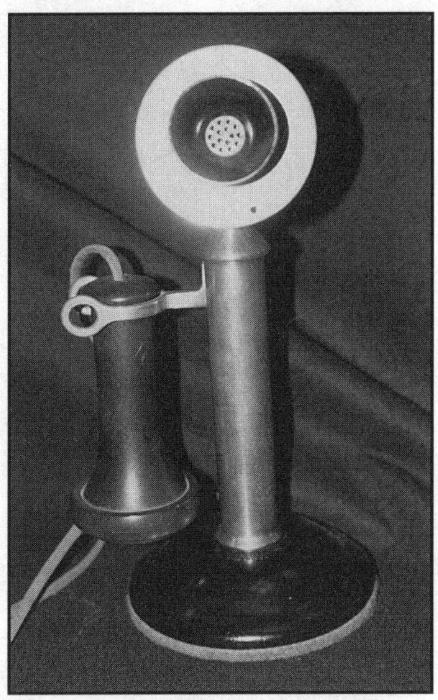

An early telephone, often called a "candlestick phone," used in the Lufkin and Conroe exchanges.

Lucy Hendrick, chief operator, far right, oversees a shift of telephone operators in the Lufkin exchange's building located on North First Street. The operators moved into the new building in January, 1953, after the conversion to a dial system.

10 Dial service comes to East Texas

In the 1950's, both Lufkin and Conroe had grown so rapidly that it became apparent to the Denman and Mantooth families that the two exchanges would have to be dramatically upgraded with new equipment to efficiently serve the communities. This realization led to changes that made the systems among the best small-town exchanges in Texas.

On May 16, 1951, the *Lufkin Daily News* reported that the Lufkin City Commission approved a two-part rate increase for the telephone company, assuring a dial system for the community. Ned D. Shands, Jr., attorney for the exchange, promised the City Commission that the dial system would be installed in 18 to 24 months and that the equipment would be housed in a new telephone company building to be built on Bremond Street, signaling the end of the telephone offices that had been housed on the second floor of the Denman Building since the company was founded.

In the previous month, according to a summary of the exchange's operations, the company had about 5,096 subscribers, up from 5,012 the previous month. Conroe had 1,842 customers in February, compared with 1,815 the previous month.

The rate increase was necessary to secure a $700,000 loan to make the changeover to a dial system. Shands told the City Commission the telephone company had been losing money on its Lufkin operations for the past few years. "The increase will put the company in the black and enable it to expand telephone service to new customers during the period the dial system is being installed," he said.

To sell its case for rate increases, the company published a brochure, "Facts about dial telephone service proposed for Lufkin." The publication pointed out: "To keep pace with this rapid growth, and provide for future expansion, the telephone company is going to offer dial service for Lufkin. But before final plans can be drawn up, the telephone company is requesting approval of a new schedule of rates that will place operations of its business here on a financial basis that will qualify it for a loan of money with which to make the conversion to a dial system. At the proposed rates, telephone service will still be one of your better bargains when you consider the higher cost of telephone materials and today's higher prices of practically all items in the family budget."

On May 25, following the second reading of the City Commission's rate increase ordinance, the telephone exchange's board of directors met in a special meeting and approved the purchase of the dial system from Stromberg-Carlson Company of Rochester, New York.

The Lufkin expansion cost the company more than $800,000, including $129,108 for a new building, $493,355 for central office equipment, $51,451 for new lines, a cable pit, and various other improvements, and $91,025 for new dial telephones.

Marion Clay, left, general manager of the Lufkin telephone exchange, and president Dr. Linwood H. Denman, center, pose with representatives of Stromberg-Carlson during the conversion of the Lufkin system to dial service in the 1950s.

On February 15, 1951, Marion Clay, a sales engineer for Stromberg-Carlson, wrote Jean Shotwell, general manager of the company, that he had finished a preliminary pricing on the dial system for Lufkin. "We have set up the pricing on the equipment, as laid out in the specifications furnished by your engineer. We can furnish our XY dial system...this is for 4,160 lines, 6,000 terminals, 12 positions at our No. 3 toll board, and three positions of D.S.A. board."

In a June 24, 1951, article, the *Lufkin Daily News* reported that "about January of 1953, the Lufkin Telephone Exchange will put into service one of the world's finest dial telephone exchanges." The new exchange, reported the newspaper, "will provide Lufkin with operator toll-line dialing between Dallas, Beaumont, Nacogdoches, Shreveport and Tyler, and will speed up long-distance calls to these points."

The newspaper added: "Lufkin has almost 6,000 phones, and the conversion of all these phones to the dial system is a tremendously complicated job. It includes the splicing of 12,272 individual wires into an equal number of wires in the new exchange. Each wire must be joined with a certain other wire. A wrong splice will mean two phones out of order. After all the wires have been correctly spliced into the dial system in the new exchange, there'll be a formal "switch-over" of all phones to the dial system."

Rex Atkinson, superintendent of the Lufkin exchange, told a reporter: "With good luck, we will be able to switch all phones over to the dial system without anyone ever noticing it...persons having phones will simply be notified to start using their dial phones." He noted that many Lufkin telephones were already equipped with dial adaptors, and all that will be required to complete the dial process would be to insert a dial in them. Other phones without dial adaptors would be replaced with new phones. The process of changing the more than 6,000 phones in town would be done to coincide with the switchover, said Atkinson.

Meanwhile, in Lufkin, as the new dial system was being planned, the company continued to deal with complaints about poor service. A letter on December 4, 1951, from Arthur Temple, Jr., then the vice-president and general manager of Southern Pine Lumber Company, focused on poor service in Diboll: "From time to time (when we have been able to reach you by telephone), we have discussed with you the matter of getting good telephone service in Diboll. Thus far, all that we have gotten is conversation (face to face, not over the phone). Inasmuch as our phone service only works about fifty percent of the time, I wonder if you would be good enough to bill

Arthur Temple, in hat, inspects dial telephone equipment in Diboll. Standing to his left are Joe C. Denman, Monroe Ivy, and Clifford Lowry. At Temple's right is Jimmie Ferguson.

us at fifty percent of the normal rate. This looks like a fair deal to me and in view of the fact that it doesn't appear that our service is going to improve, perhaps we could get our rate lowered as the only other alternative."

Temple's letter coincided with a petition from Diboll and Burke subscribers protesting a 15-cent charge for calls between the two communities.

Six days later, Kester W. Denman, Sr., of the telephone company replied to Temple's letter. "Lufkin Telephone Exchange admits that the Southern Pine Lumber Company and the people generally in and around Diboll are entitled to as good telephone service as it is humanly possible for us to give, and it is our intention to do everything within reason to furnish good service. It is trite to say that Rome was not built in a day, but it is still true." Denman explained to Temple the improvements being made in Lufkin and promised that Diboll improvements will follow. He concluded with: "The Board of Directors does not deem it wise at this time to either lower your rates or cut your bill to fifty percent of the normal rate."

In 1951, the Lufkin exchange board was made up of president L.H. Denman, first vice-president N.D. Shands Jr., second vice-president H.T. Bowers, acting secretary B.H. Denman, and Kester W. Denman, Sr., and Peyton R. Denman. Jean Shotwell continued as general manager and secretary-treasurer. The same board governed the Conroe exchange, but Morris Bateman had become the manager of the Conroe operations.

In the November, 1951, telephone directory, black customers were identified with the notation "col. (for colored)" beside their name. In the yellow pages, funeral homes were distinguished as being white and colored. Residential listings were distinguished from businesses with an "r."

As 1951 drew to a close, the company's employment rolls in Lufkin included operators Charline Albritton, Sue Burkett, Maxine Carpenter, Elizabeth Clampit, Dorothy Ellis, Dorothy Fletcher, Irene Flurry, Joy Graham, Helen Graham, Peggy Hataway, Katherine Hayes, Lucy Hendrick, Danie Horton, Ora Lee Jones, Reba Jones, Lera Dell Martin, Doreen Massingill, Betty Mathews, Maurice McCoy, Archie McGee, Mary Modisette, Maxine Parrish, Jeweline Redd, Alice Register, Mabel Rhodes, Ouida Joe Smith, Marjorie Solley, Lois Stanaley, Nela Sullivan, Ruby Swor, Sue Tolbert, Faye Nell Teer, Joyce White, and Ima Jean Wilkerson. Office employees included Louise Anderson, Nina Brookshire, Winona Etheridge, Christine Herrington, Mozelle Hopper, and Marie Lankford. Avis Maas was the bookkeeper, assisted by Acacia McMullen. Betty Warren was a secretary. Linemen were A.J. Barron, Victor Bass, Bertis Brock, M.E. Burrous, A.S. Carrington, M.L. Hand, Seburn Innerarity, Monroe Ivy, Charlie Lawrence, Jack Mott, Samuel Posey, Obie Read, E.C. Weaver, and Sam Williams.

Marion Clay, left, explains the operations of Lufkin's new dial equipment in the 1950s to Bill Collins, center, and Harold Schmitzer.

Conroe employees included operators Lucinda Adams, Betty Brinkley, Dorothy Carter, Juanita Castleschouldt, Juanice Coleman, Mazelle Doughty, Evelene Eddins, Betty Gamble, Eunice Hill, Lanell Hodges, Betty Jo Holladay, Lilla Jeffcote, Jewell Kelley, Janice Lowe, Reba Mizell, Mary McNulty, Betty Joe Patterson, Rosa Lee Poland, Helen Toews, Francis Turner,

Marion Clay, right, poses with Jack Mott, Durrell Gray, and Morris Bateman during an equipment inspection at Conroe in 1958.

Bobbye Williams, Nellodyne Presley, Margie Price, and Ivy Wagers. Office personnel were Doris Austin, Hattie Duran, and Mattie Wilson. Linemen were Carl Beeson, Bob Ellisor, and Fritz Pearson.

In March of 1951, Mrs. Mary Caroline (Biddy) Denman, 91, widow of Dr. Alexander Madison Denman, passed away at her Lufkin residence. She was buried beside her husband's grave in Glendale Cemetery. Mrs. Denman, the daughter of Thomas R. Walker and Emily Z. Briscoe, was a charter member of the First Baptist Church, a member of the Order of the Eastern Star, and the Woodman Circle.

Jean Abbott, a company retiree, recalls that during the mid-1950s, the Lufkin telephone business office manager, Lila Baird, was in complete control over the office, and her office workers were prohibited from removing a charge from a customer's account without her approval. However, a physically handicapped college student who was having a difficult time paying his account was disconnected several times for a lack of payment. When he came in to pay his account, no one in the office had the heart to tell him he owed an additional two dollars for reconnecting his telephone. So several employees would pool their money to pay the reconnect charge without telling the business office manager.

In those days, all service orders for telephone orders were manually written. When the orders were completed, they were returned to the bookkeeper, who compiled the number of days to bill the customer and the partial month on a ten-key adding machine, recalls Mrs. Abbott. The office lacked a copy machine, so carbon paper was used to create copies on a manual typewriter.

The customer's name, telephone number, and address were cut into a small metal plate on a Graphotype machine and printed into the statement on an Addressograph machine before billing. The statements were then entered into an NCR billing machine by a bookkeeper, who punched in the exchange service, tolls, directory advertising, installation charges and previous balance. The statements were billed in six cycles every five days for Lufkin. Conroe had three billing cycles every

The Conroe Telephone Exchange moved into this building on Phillips Street in 1954, ending 25 years of operations in the Madeley Building.

10 days.

"The Lufkin business office staff increased to two bookkeepers in 1954 and that brought the entire office staff to 12 people, including the managers," said Mrs. Abbott. "The office manager and general manager were the entire accounting personnel, personnel, separations, accounts payable, rates and tariffs and accounts receivable. It was unbelievable that only 12 people could do all of these functions, but Lufkin had only 5,000 customers at the time."

With the arrival of 1952, Marion Clay's work with the Lufkin exchange as Stromberg-Carlson's representative on the Lufkin dial conversion project impressed the Denman and Mantooth families so much that he was hired to become the first general manager who wasn't a family member. Born November 6, 1906, in Tom Green County, Texas, Clay came to Lufkin from San Angelo and came from a telephone family. He was a graduate of Texas A&M University.

"My father was in the telephone business for 48 years, so my brother and I started in the field of telephony at a very early age," he wrote on an LCTX biographical form in 1979. "I started working Saturdays and summers climbing in the construction crew. My records go back to 1926 with the San Angelo Telephone Company. I worked in construction, cable, installation and repair. During this early time, I also worked on magneto equipment, common battery equipment, and dial equipment, as well as all types of PBX equipment and toll boards. In 1928, San Angelo purchased Automatic Electric Dial Equipment, and after completing a special dial school at Automatic Electric in Chicago, I was made equipment engineer and assistant superintendent and helped with the installation of the San Angelo dial equipment in the fall of 1928."

From 1928 to 1945, Clay made all additions to the San Angelo system, engineered and installed telephone equipment at Goodfellow Air Force Base and Concho Bomber Base at San Angelo during World War II. "The draft board decided I was needed more at the Air Force bases," he said.

After the war, Clay joined the Stromberg-Carlson company in Rochester, New York, and after undergoing special training, began a new job as a sales engineer in Stromberg-Carlson branch office in Kansas City, Missouri, serving 11 states.

Accepting the Lufkin Telephone Exchange's offer of employment in 1951, Clay moved his family to Lufkin and started working for the company April 4, 1952.

Clay, who died September 9, 1982, is remembered by long-time employee Sue Roberts as "a father figure who had a remarkable ability to listen patiently to people." Roberts added: "He understood the company's customers, and I never saw him mad in my life. He was always smiling."

Clay had reason to smile on Saturday, January 24, 1953. He was on hand for the exchange's dial conversion at midnight that day.

At the conversion ceremony, Mrs. M.S. Townsend, who had phone Number 3 under the old manual telephone system, made the first dial call on the new system to Lufkin Mayor Granville Wright. Mrs. Townsend had had the same telephone number since 1900 when the first switchboard was installed in the town.

The company decided to make the dial conversion at midnight when few people were using their phones. However, many Lufkinites stayed up so they could use their new dial telephones.

The introduction of dial telephones in Lufkin brought an end to a unique practice by the telephone company. Until 1953, the company had kept a "Redlight Book," in which clerks penciled in the names of people who had not paid their bills. If a customer failed to pay, operators switched on a red light above the customer's number on the switchboard and declined to answer calls originating from the number. However, she always placed incoming calls to the number in case the call was of an emergency nature.

Numbered holes on the switchboard were also occasionally plugged if phones were taken out of service. These were called "takeouts."

Sue Roberts remembers another practice in Lufkin. "Before we left the old Denman Building in the 1950's, several of the girls had been assigned to walk around town and collect money from pay phones in various buildings. One of the pay phones was located in the Bonner Hotel, which was known as a house of ill repute," she recalls. "The girls who had to pick up this money would wait until the streets were empty of pedestrians and traffic, run into the hotel and get the money, and then wait again until the streets were empty before leaving the building."

Mrs. Roberts recalls that when the company's new building was built, it didn't have a place for the employees to take coffee breaks, so workers walked four blocks to the corner drug store to have their coffee.

One of Betty Warren's first jobs after the dial conversion was to assign new telephone numbers for everyone in the exchange. "The old system had only three digits, plus a letter like a J, and the new numbers had six digits like NE 4433." The letters represented the name of the exchange, i.e. NEptune.

"All telephone bills were to be paid by the tenth of each month. If they were not paid, the office girls would call each customer to remind them," recalls Mrs. Roberts. "If a bill still had not been paid by another six days, a second call was made, and an installer would go to the customer's house and collect the money. Disconnection was the final step."

The Lufkin dial conversion also enabled Clay and his staff to provide better PBX service and in 1954 put into service a 160-line PBX for Texas Foundries, one of the city's largest industries. Clay reported on the equipment to a reporter: "The need for rapid, economical 24-hour communications within the plant, as well as two-way trunk service to Lufkin's XY dial central office, led to the selection of the equipment."

With the Lufkin conversion practically complete in 1953, one of Clay's first assignments as the general manger of the Lufkin and Conroe systems was to launch an improvement program in Conroe similar to the one he had engineered in Lufkin.

On May 12, 1952, Kester W. Denman, Sr., wrote Bill Newton, the Conroe mayor who had been a vocal critic of the Conroe telephone system. "The Board of Directors of the Conroe Telephone Company, at its regular meeting...May 10, 1952, unanimously voted to convert the present manual telephone exchange at Conroe to dial, and to open negotiations at once with financial agents, building contractors, and telephone equipment manufacturers for the purpose of obtaining funds, erecting a building, and purchasing the equipment."

Buff and Phelps, a Chicago public utility analyst firm hired to evaluate the company's needs in Conroe, reported to the board of directors: "...Conroe has experienced very substantial population growth over the past four decades and...the outlook is for continued favorable growth for some time to come."

The firm reported that "the company's central office equipment is in reasonably satisfactory condition although quite inadequate." In April, the company reported 2,352 subscribers, compared with 2,339 the previous month, reflecting a steady growth trend.

A story in *The Houston Post* said the system would cost $500,000, including a $100,000 building on Phillips Street, two blocks from the courthouse. At the time, the exchange had 1,600 subscribers, double the total from five years earlier.

In an interview with the *Conroe Courier*, Denman said "an increase (in telephone rates) would be inevitable," but he was unable to estimate the extent of the rates. It was pointed out by Denman that Conroe residents enjoyed cheaper phone rates than any of the towns surrounding it.

In January, 1953, Clay appeared before the Conroe City Council to formally ask for a rate increase to support the expansion. Clay said the rates he proposed would remain among the lowest in East Texas. A one-party residential rate for dial service in Conroe was to be $5.25, compared with $7.75 in Navasota, $10.00 for Brenham, and $8.00 in Liberty and Columbus.

Despite Clay's assurances, however, the Conroe City Council was reluctant to approve the new rates and ordered a public hearing on the issue. The council claimed the rates were higher than new rates recently put into place at Lufkin before a dial system changeover.

When the hearing was held in February, more than 50 Conroe residents showed up, but were divided in their sentiment toward the telephone company.

The rate issue continued to be debated in the community until April 13 when the Council approved a series of interim rates allowing the telephone company to proceed with the installation of a dial system. The rates approved by the council were below those originally proposed by Clay in January, leading the telephone company to reject the Council's compromise. In a letter to the Council, Dr. L.H. Denman wrote: "We have given the proposed rates a serious study, and find they will not justify the investment required for the conversion program."

A month later, the Council accepted a new offer by the telephone company and put into effect a schedule of interim rates, followed by another schedule that would be allowed when the dial system was installed. On May 31, the *Conroe Courier* published a bulletin on its front page: "Marion Clay, general manager of the Conroe Telephone Company, said Tuesday...that the company is trying to get all equipment contracts closed this week and is pushing the work necessary...to install dial switchboards. If all arrangements can be completed this week, the

The Lufkin exchange's plant department posed for this photograph in 1959. Standing, left to right, are Jack Mott, Joe Wadsworth, Gerald Glover, David Colwell, Kenneth Calhoun, Dalton Conner, Obie Read, Johnnie Weatherford, unknown, Joe Noel, Melvin Burrous, Marvin Schmidt, Robert Beavers, Robert Brock, and Doug Shands. Kneeling, left to right, are Monroe Ivy, Norman Tindall, Gus Brewer, Maxie Pearson, Bobby Coleman, Ike Smith, Clifford Lowry, Tom Dominey, Walter Lawrence, Billy Edwards, and A.S. Carrington.

interim rate will be added on June 1 telephone bills, which will be for June service. Otherwise, the interim rate will not be added until July 1 billing, he said." Stromberg-Carlson had been chosen to install the dial system, Clay said.

Meanwhile, with the completion of the Lufkin dial system, Clay turned to the problem of poor telephone service in Diboll.

On Saturday, December 14, 1953, more than 400 telephones in Diboll were converted to dial usage, culminating a promise by Lufkin officials to give Diboll improved service. Preceding the changeover, Lufkin officials joined Diboll residents in a celebratory dinner at the Pine Bough Restaurant. The 30 people in attendance heard brief remarks from L.H. Denman and Marion Clay, Arthur Temple Jr. of Southern Pine Lumber Company, and County Commissioner Bill Collins.

The new Diboll system enabled residents of the community to make calls to Lufkin without being charged a toll fee. They were also charged the same monthly rates as Lufkin subscribers. The controversial 15-cent toll fee between Diboll and Burke was also eliminated.

While the Lufkin telephone system used the NEptune prefix, Diboll phones were preceded with a MYrtle prefix.

There were special provisions, however, for using the two prefixes. A party line user wanting to contact someone on his or her own party line had to first dial 7, hanging up to give the other party enough time to answer, and then lifting the receiver to begin the conversation.

A telephone subscriber in Lufkin wanting to speak with a party in Diboll had to dial the MYrtle exchange number. The same procedure applied to someone in Diboll wanting to dial a number in Lufkin's NEptune exchange. Lufkin and Diboll residents calling someone in their own exchanges, however, did not have to dial the exchange prefix.

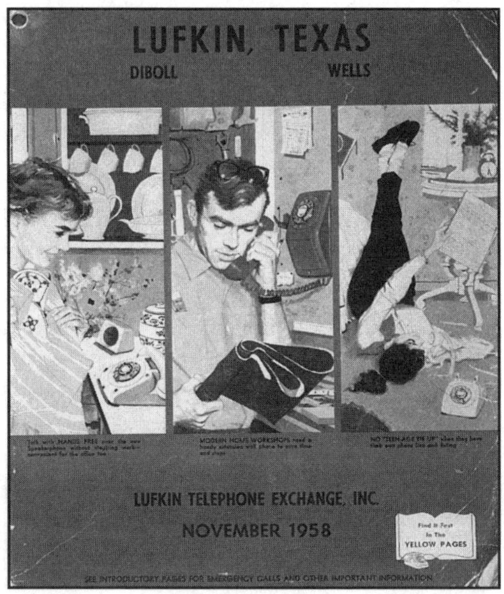

Lufkin's telephone directory in 1958.

Once the Diboll exchange was converted to dial, eight Lufkin telephone crews spent their Sunday in the community, making routine changes brought on by the equipment conversion. Clay promised Dibollians the followup work would continue until all problems were worked out.

"The equipment and work done in this project has cost us about $127,000, but completion of this project is the completion of a dream we've had for some time," Clay told the Pine Bough dinner group. "Our equipment in the Lufkin and Diboll exchanges is the best available anywhere," said Clay. "We can put up to 10,000 sets in operation here in Diboll without adding any equipment. Too, both cities are now set up for station to station long distance dialing by the individual without going through the operator.

Morris Bateman remembers when the dial system came to Conroe:

"We didn't have the money to buy the system, so we went to Stromberg-Carlson itself, and borrowed $350,000, and they agreed to install the system. Meanwhile, we hired Jack Bankston to build a new exchange building on Phillips Street. While Stromberg-Carlson was building the equipment in Rochester, New

York, we were working on the building, and at the stroke of midnight on Saturday, September 25, 1954, we had a cutover night with every employee of the company, along with Marion Clay (the company's general manager) and a lot of the folks from Lufkin. At one minute past midnight, we turned on the new system, cut off the lines to the old switchboard, and after that, every time someone dialed a phone, the equipment did all the work. The days of the switchboard operators were over in Conroe, and everybody in town wanted to use their phones to see if the dial system really worked."

In November of 1954, *Telephone Engineering and Management*, a trade journal, reported that the Conroe system—which included 2,000 line terminals, 3,000 connection terminals, and 25 toll circuits directly into Houston—qualified Conroe as the biggest central dial office in Texas.

The Conroe dial system enabled the exchange to handle more than 2,500 calls at one time, which essentially solved most of the company's problems in Conroe.

Like most dial systems of the day, Conroe also had an exchange name, PLeasant. Later, as the telephone systems accumulated more customers, the prefix letters were replaced with seven-digit numbers.

"People all over town were excited about the new dial system," recalls retired Conroe employee Doris Johns of Lufkin. "They were finally getting the kind of service available down the road in Houston," she recalls. "All of our employees were excited, too, because we were finally able to move out of our cramped second-floor office in downtown Conroe to a nice new building on Phillips Street."

The dial system came just in time for Conroe. The town's growth began to mushroom with the completion of the new Houston Intercontinental Airport, less than 40 miles down Interstate 45, and Lake Conroe, which attracted tens of thousands of new residents to Montgomery County.

When dial service came to Montgomery, the employees of the company actually installed the dial equipment in a building they constructed in Conroe during their spare time. Once the equipment was installed in the completed building, the entire exchange was transported by a house mover to Montgomery and the actual cutover was made several days later at 7 a.m. in the morning.

Members of the Lufkin and Conroe boards from 1952 to 1960 included L.H. Denman, N.D. Shands, Jr., Kester W. Denman, Sr., H.T. Bowers, Peyton R. Denman, B.H. Denman, Thomas O. Shelton, Jr., Frank R. Denman, and D.C. Hucherson. The Lufkin Telephone Company became the Lufkin Telephone Exchange in 1957, and N.D. Shands, Jr., joined the board as first vice-president.

Telephone operators at work in the 1950s at Conroe. Left to right are Eveline Eddins, Lilla Jeffcoat, Vera Acrey, Helen Towes, Reba Mizell, Eunice Hill, Lenell Hodges, and Virginia Waller, at the desk in right foreground.

11 The end of an era

In 1953 and 1954, the era of the centralized telephone switchboard came to an end in Lufkin and Conroe. The operators who had directed telephone calls in the two communities for a half-century were soon replaced by dial telephones.

Few of the operators, however, lost their jobs with the telephone company. The company remained essentially a family business, so the operators were placed in other positions within the company. "I think that's what made the company so successful," remembers long-time employee Sue Roberts. "The employees were always looked upon as members of the family, and they were treated that way. In turn, they put their hearts into their work."

All of the company's switchboard operators at Lufkin started with the exchange while it was housed on the second floor of the Denman Building on South First Street. They remained there until 1953 when the company built a new building a few blocks north of the Denman Building.

All of the operators who worked in that era have since retired, but remember their switchboard days with fondness.

Lucy Hendrick of Lufkin, who retired in 1980 after 36 years with the company, started as a telephone operator working under chief operator Helen Campbell, a member of the Mantooth family often referred to as "Aunt Helen."

"In those days, all the girls on the switchboard worked for 15 cents an hour, and were glad to get the money because times were hard and jobs were difficult to find," recalled Mrs. Hendrick. "We were paid with cash money inside individual envelopes, but when one of the girls left her money in the lounge and it turned up missing, the company started paying us with checks."

Lucy Hendricks, chief telephone operator, and Lila Baird, business office manager, stand in front of the new telephone offices built at Lufkin in the 1950s.

To operate the switchboard—which Mrs. Hendrick compared to an upright piano filled with small holes—the operators sat before clusters of holes, small lights and cords.

Each operator's position included a board space about 30 inches wide and 40 inches deep with a small light under each number hole on the board. When a subscriber's telephone receiver was lifted from its cradle, the light under the corresponding number hole would light on the board. An operator would respond with the words, "Number please," and then connect a cord from the lighted number to the requested number.

The numbers on the switchboard were grouped in blocks, beginning with 200 and ending with a 3,000 series.

"Most people in town knew the numbers they wanted to call, but if someone couldn't remember a number, or didn't have a directory, the operators switched them over to Mrs. Campbell, who doubled as the directory operator," said Mrs. Hendrick.

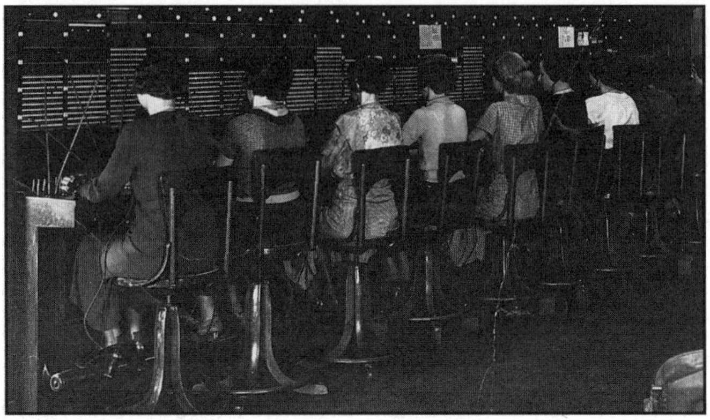

Early switchboard operators at work in the Lufkin telephone office, located above a downtown drug store.

The operators who manned the switchboards came to work in "nice dresses," recalled Mrs. Hendrick, "but because many of them often had dates after work, they brought with them a work dress, and changed back into their nice dresses when they finished their shift."

Mrs. Hendrick remembers Mrs. Campbell as "very strict, but very fair and always willing to listen to our problems." Mrs. Hendrick recalls that Mrs. Campbell had a "sneak board" which enabled her to occasionally monitor conversations. "We weren't allowed to carry on conversations with our friends or relatives; that was a strict rule because we were supposed to be professional in our jobs," said Mrs. Hendrick.

One of Mrs. Hendrick's jobs was to assist Mrs. Campbell in the preparation of the company's telephone directory. "In the early days, there wasn't much to the directory, but I had to call off all the numbers several times a day to make sure they were all correct," she remembers. "We worked on it for a week, and I became hoarse, and could barely speak."

Chief operator Lucy Hendrick, at desk, and nine operators at Lufkin. Left to right are Evelyn Free, Rose Baker, Melba Rudd, Jeanette Davidson, Danie Horton, Maxine Gaston, Lanell Hodges, Jeweline Redd, Ora Lee McPherson, Ima Jean Anderson.

Mrs. Hendrick was promoted to Mrs. Campbell's assistant and a few months later, general manager Jean Shotwell came to the switchboard room and told Mrs. Hendrick that Mrs. Campbell would not be returning to work after an absence of several weeks. "I was making 15 cents an hour in those days, and they gave me a nickel raise," she recalls.

Mrs. Hendrick said she "always enjoyed working" for the company. "Times were hard, but I loved the people, and we were always happy," she remembers. "We felt like we were a family, and we stuck together.

Ivaline Parrott was one of the few telephone operators who didn't work for the

Lufkin exchange, but she worked side-by-side with Lufkin's local operators. Fresh out of Lufkin High School, she was hired as an operator for Southwestern Bell Telephone Company.

An early telephone company dinner at the Angelina Hotel in Lufkin, presided over by president Dr. Linwood H. Denman, front left.

"Both Southwestern Bell and the Lufkin Telephone Exchange were side by side, on the second floor of the Denman Building at the corner of Dozier Avenue (now Frank Avenue) and North First Street, just above the Lufkin Drug Store," she remembered in an article written by Lufkin author Margie Kirkland. "All the telephone operators worked from one long board, about three and a half feet high, from the sitting position, with the Lufkin Telephone Exchange's local operators on the left and Southwestern Bell's long-distance operators on the right. There were ten girls who handled local calls and five who serviced long-distance calls."

Mrs. Hall Weaver was the chief operator supervisor for Southwestern Bell. Operators handling local calls had their own supervisor. The business office for the Lufkin exchange was on another part of the second floor.

"There was one headset for each operator position," remembered Mrs. Parrott. "The earpiece was placed over the right ear with a wire band across the top of the head to hold it in place, and we had a horn-type mouthpiece resting on the chest. Each time the operator changed, the mouthpiece was disinfected with wintergreen."

The telephone office worked around the clock, and Mrs. Parrott worked a split shift, from 3 p.m. to 11 p.m. or 5 p.m. to 9 p.m., or 9 a.m. to 1 p.m., then off until the 5 p.m. shift. Two local operators and one long-distance operator usually worked the night shift.

If a local operator answered a call and the caller wanted to place a long-distance call, the operator would transfer the call to a long-distance operator in the same room. If the caller wanted to reach a telephone in a city where there were no direct lines, operators had to route the calls through several cities to make the connection. The caller on the line would hear operators requesting destinations and numbers to complete the call.

Completing calls from rural areas was a little more complicated. "Rural lines had several telephone lines on each line with different rings for each subscriber. Callers would place their calls by stating the desired number, along with the number of rings, such as two shorts and one long. All telephones on the rural line would ring, but only the party with the specified number of rings was supposed to answer the call.

Even though it was difficult to avoid listening to conversations on the telephone lines, telephone operators adhered to a strict code of silence. Mrs. Parrott remembers that Mrs. Weaver, a Sunday School teacher, often told her operators it would be a sin to tell what they heard on the job.

Mrs. Parrott remembers that operators in the 1940s and 1950s were considered professionals and were paid well. Their paychecks averaged $150 to $200 a month, which was considered good pay. Supervisors kept up with the operators' working hours by keeping time cards on each employee. "All of us wore dresses in those days," remembered Mrs. Parrott, "because when you had a job as nice as ours, you wanted to dress up pretty to go to work."

Lufkin telephone company employees, left to right, Dorothy Spivey, Jean Abbott, Nancy Kirkland, Carlon Reynolds, Sue Roberts, Barbara Davis, and Margie Tindall. This photo was taken in the early 1950s.

Like most operators, Mrs. Parrott walked to work, a distance of about four blocks. Leaving her home on East Dozier, she walked past Atkinson Candy Company, International Harvester, D.O. Wallis' grocery store, the E.W. Leach Grocery Company, Moore Grocery Company, Henderson-Medford Motor Company, the Double Dip Ice Cream Parlor, the county courthouse and jail, Hubbard and Melton Grocery, the Lynn and Texan Theaters, the Yellow Cab Company, the Eat-A-Bite Cafe, and finally to the Denman Building for a climb up the steep stairway that led to her workplace.

"I was never afraid to walk to work or back home by myself," she recalled. "Everyone walked; it was the lifestyle then. And no matter what the hour, our house was always unlocked. Like our neighbors, our family never gave a thought to being burglarized."

Even though the duties of the telephone operators changed after the conversion to dial systems, some operators were still essential for certain functions. Edna (Koochie) Carlisle, who began work with the company in 1957, remembers her work as a post-dial operator.

"There were about 80 people employed in traffic (long distance and directory assistance). We had huge directories where the names, addresses and telephone numbers were handwritten. We got the orders (for local numbers) from Sue Roberts in the business office, and wrote them in the directories," she said.

"On the long-distance side, the first position answered calls to the business office and calls from operators in other towns. These were called inward calls. Inward was for connecting other operators to our non-dialable places that were ring-downs and any calls with difficulty getting through," she said.

Miss Carlisle recalls that operator position No. 2 was called "Operator 21" and handled all delayed calls that went through the Dallas area. Position No. 4 was called "Operator 41" and went through the Houston area. Position 3 helped the other two operators take care of delayed calls.

If a customer did not complete a call, and wanted the operator to continue trying, the operator would advise the customer she was placing the call with operators 21 or 41. The customer was asked to call back and ask for 21 or 41.

"Operators 21 and 41 always had a large stack of delayed calls, some of them a week old. We had many, many calls to the military, and we had to call the marine operator for calls to ships and the overseas operators for calls to foreign military bases," she said.

All the tickets for long-distance calls were handwritten. "When I first came to work for the company, we used prefixes like CApitol or JAckson, writing the first two letters on regular calls, and the complete word on collect calls. On all calls, we wrote the customer's name. Operators at the directory assistance positions hand-rated all of the tickets by using a ratebook."

Operators working in the exchange had cards clipped on their headsets designating their seating positions. "This way, we spent time at all the job responsibilities, and we did not leave until someone took our place," recalls Miss Carlisle.

Operators had three different shifts—days, evenings, and split shifts. They also found they had different types of customers. "During the day, we had mostly business calls. At night, the calls were mostly from residential or public telephones. If you worked a split shift, you had both.

During Miss Carlisle's early years, calls to Diboll, Wells and Alto were long-distance charges. "Wells' telephone office was inside the Woods' house (the Woods family owned the local exchange) and in Apple Springs, there was only one telephone,which was in a drug store owned by Mrs. Byrd Evans, the mother of a Lufkin physician. If a customer left owing money, Mrs. Evans rounded them up and they paid off. These were called toll stations, and there were many places like these all over the country. We also had ring-downs in places like Wells, Alto and Huntington," she said. Ring-down locations had an operator who completed the call, mostly to party lines with numbers such as 241J or 241R.

In 1957, many towns had their own operators, just as Lufkin did, and Miss Carlisle recalls that operators could tell the name of the town by the particular ring emitted by the operators' equipment. "I remember the day Crockett cut over with its new equipment, and I had a customer who refused to believe I was dialing the right number because he was accustomed to a certain sound," said Miss Carlisle.

In the late 1950's, the traffic department where Miss Carlisle worked was on the first floor of the Lufkin Telephone Exchange's new building at the corner of Bremond and North First Street. "The building did not extend toward Angelina Street as far as it does today (in the 1990s). There was also an old hotel, the Mahaffey Hotel, which was more of a rooming house, about where the drive-in lane is now located. Some of us often walked down to the drug store, which was located on the first floor of the Denman Building, when we needed a break, or we

Shown in the Lufkin customer service department in the 1950s are, in the front, Carlon Reynolds and Patsy Giles. Left to right on the second row are Sue Roberts, Billie Lewis, and Jean Abbott. On the third row, left to right, are Bobbie McGilvra, Peggy Davidson, and Lucille Stroud. In the back is Katherine Harris Gay.

could walk down into the basement of the exchange building, where there were chairs and some couches," she said.

Miss Carlisle graduated from Lufkin High School in May of 1957 and had worked at several jobs, but wanted full-time work. "I went all over Lufkin applying for a job. I started home in my 1953 Ford and remembered that I had not gone to the telephone company. I started to drive on home, but I turned around and went to the telephone office and filled out an application. The last week in November, Lucy Hendrick, the chief operator, called me and asked me to come to work. I was shocked and thrilled," she recalls.

"Working at the telephone company, I made more money than anywhere else I could have worked, and as the years went by, I always had a good job. The company always wanted to make a profit, but wanted the people who worked there to make a profit, too. I have my biological family, my church family, and I have my Lufkin-Conroe family. If I had not turned that old 1953 Ford around and gone back, I would have missed out on one of the greatest families I could ever know," she said.

Vera Cotton Acrey of Conroe, who worked as an operator in the old Madeley building in downtown Conroe, came to the switchboard as a part-time employee and later graduated to a full-time job.

"The phone company was located on the second floor of the Madeley building with a dress shop below. To go to work, we'd walk up a steep flight of stairs. At the top was a pay phone and to the left was the business office, where Doris Austin and Mattie Mae Wilson worked. To the right was a six-by-six area which was used as a lounge for the operators, and beyond that was the switchboard room, where the company had five local service stations and two long distance stations," Mrs. Acrey recalls.

A telephone of this type was used in Conroe from 1940 to 1954.

During her service as an operator between 1952 and 1954, a tarp was hung over the switchboard. "The roof leaked every time it rained, so Mr. (Morris) Bateman put the tarp over the board to keep it from getting wet," she said. "But the rain ran on the floor and fell on the dress shop downstairs."

As a high school student, Mrs. Acrey worked part-time from 5 to 9 p.m. on weekdays and all day on Saturday and Sunday. During the summer of 1952, she worked alone at night. "I stayed scared most of the time when I worked at night. One night, I kept hearing strange noises in the room with me. I called Mr. Bateman, who came over. He found some big old rats in the room," she said.

Conroe in the 1950s was a small, quiet community and few business firms remained open at night. An exception was Hulon's Gulf Service Station, and Vera established a running telephone friendship with Calvin Blake, the station's night manager. "Both of us were sleepy, so we'd call each other to stay awake," she said.

Telephone operators in the early 1950s were called upon to provide more than calling services. One service was a primitive method of call forwarding. "There were a couple of businessmen who didn't have secretaries, so when they went out for coffee, they'd call us and tell us to refer their calls to the Capital Drug Store," she recalled.

In 1954, when dial service came to Conroe, the switchboard positions were eliminated, but the company offered all of the operators new employment in Lufkin. Mrs. Acrey, however, decided to remain in Conroe and lost her position. However, in 1962, the company offered her a new job in the business office at the company's new Phillips Street building in Conroe. One of her responsibilities was to maintain the Conroe telephone directory, assuring its accuracy, and working with the company that published the directory.

In 1996, when Mrs. Acrey retired after spending nearly 35 years with the Conroe exchange, the company presented her with a series of gifts—including the chair she used for the entire 35 years in the business office.

"An office supplier from Lufkin brought the chair to Conroe in 1962 when I returned to work," she recalled. "I was offered new chairs, but I always liked my old one, so I stuck with it. Every now and then, the man from Lufkin would drive to Conroe to adjust the chair or make a repair, but it always worked."

When she retired, Mrs. Acrey achieved another distinction. She was the only company employee to work at all five of the company's Conroe locations: the Madeley building, the Phillips Street location, the Davis Street location, the North Frazier address, and finally the present-day office on the South Loop.

By the 1960s, dial telephone equipment installed at Lufkin and Conroe had ended many of the problems encountered by the Lufkin and Conroe exchanges. Here, John Harris, right, and Jack Mott, center, of the Lufkin exchange visit with a Mr. White of the Bell system in the Lufkin equipment room.

12 The sixties: an improved reputation

With modern dial systems in place, the Lufkin and Conroe telephone systems owned by the Denman and Mantooth families entered the 1960s with an enhanced reputation for customer service.

Much of the credit went to Marion Clay, a warm and friendly West Texas telephone engineer who moved to Lufkin in 1952 to become general manager of the Lufkin-Conroe Telephone Exchange.

Long-time employee Sue Roberts remembers Clay as "a warm, pleasant man who never seemed to be upset with anyone or anything." Lucy Hendrick, a former chief operator, recalls him as "the kind of man you could always talk with."

Clay also had a broad reputation outside Lufkin and East Texas. He became a director of the Texas Telephone Association a year after coming to Lufkin and served continuously on the board until 1973. He became second vice-president of the association in 1956, served as first vice-president for 1957 through 1960, and became president of the state organization in 1960, serving through 1968. He had also served as chairman of virtually all TTA committees during his span of service.

Clay was also elected to the board of directors of the United States Independent Telephone Association in 1960 and served until the fall of 1963. He left the board that year to allow the new president of the Texas association to serve on the national board.

During his career, Clay became a member of the Independent Telephone Pioneer Association, a group formed to honor telephone leaders, and became a life member. He was also a member and president of the Pineywoods Pioneer Club.

Buoyed by new growth in Lufkin and Conroe and the success of the new dial systems installed in the early 1950s, the two cities' telephone exchanges began to grow by leaps and bounds in the 1960s.

Marion Clay, who led the Lufkin and Conroe telephone exchanges in the 1960s.

In February, 1961, death claimed Jean Shotwell, 61, the former manager of the Lufkin-Conroe exchanges. Shotwell was a member of the First Christian Church, past district governor of the Lions Club, a member of the Salvation Army Advisory Board, and active Red Cross worker.

By March of 1961, the Lufkin telephone exchange had grown to 10,435 subscribers, a doubling of new telephones since 1951, and the Conroe system had grown to 4,130 subscribers—an increase of more than 2,000 new customers in 10 years.

A Lufkin telephone directory in 1963.

The year also marked a milestone of sorts for the exchange. The company, which had aggressively pursued the PBX business in Lufkin, installed at Perry Brothers' downtown Lufkin headquarters a new Stromberg-Carlson F-50 PBX system. The installation was the first in the Southwest and one of the first in the United States. The installation was made entirely within a portion of an unused elevator shaft and initially operated 23 stations in two Perrys locations in Lufkin. Perry Brothers, which operated 85 variety stores in Texas and Louisiana in the sixties, closed all of its stores in 1997.

In 1963, Clay had the privilege of presenting State Senator Martin Dies, Jr., of Lufkin a special telephone during Dies' service as governor for a day while Governor John B. Connally and Lt. Governor Preston Smith were out of the state. Clay headed a Lufkin delegation that traveled to Austin for the Dies ceremonies and to present the telephone, which carried the State of Texas seal and a special engraving to commemorate Dies' service. Joining Clay during the ceremonies were W.S. Bain of San Antonio, president of the Texas Telephone Association, and J.B. (Jack) Haley of Austin, executive vice-president of TTA.

The year also brought an end to a property dispute between Buna and William N. Lambing and the Lufkin Telephone Exchange. The dispute arose over a boundary between the Lambings and the exchange. The dispute had been going on since 1953 and involved property between the exchange's new building on North First Street and its next-door neighbor, the Mahaffey Hotel and boarding house.

In 1967, the Conroe exchange placed a newspaper ad in *The Conroe Courier* and shared with its customers a progress report entitled, "1967 AND YOUR TELEPHONE."

"The Telephone Company is now placing and splicing its large cables in the underground duct system whereby it can get new cables from the office out into the streets. The existing poles wouldn't hold up any more new cables, so time and money had to be spent for the construction and placement of an underground duct system. Termination terminals are being installed to connect 1,800 pairs in new cables for new lines out in outlying areas. The central office equipment to work these new lines is already in place and working.

"A new type rural board is now working, which eliminates any telephone bell from ringing except the one being called. New telephones with bells that will work with this board are now on order, so we can change every existing rural set in the country.

"By May 20, we will have 44 trunk lines to Houston on the operator O level. Also, 20 lines to Houston on the direct distance dialing 1 level. This company has had this equipment ready for use since November of last year, but there were no new lines into Houston for same to work.

"Since March 1st, we have been testing and allowing the phones in Conroe and Montgomery to dial each other direct free of charge. Announcement and instructions have been published on this additional service.

"An order has been placed with the Purgas Company for a machine to be installed to keep the large cables going out of the central office dry. It blows air into them, therefore keeping out the water, which now puts out of order the paper-wrapped wires under the lead. It will be added to the older cables as soon as the workmen can block the terminals. All new cables that we buy now are made out of the plastic-wrap and are not subject to moisture.

"These are just a few of the things in the making for this year, and more are on the drawing board for the future. As you would expect, all of these services have cost large sums of private and borrowed capital, which this company has raised at no extra cost to its subscribers. Money for construction and installation of this type comes from your rental of telephone services only, but the requirements for outside and inside the plant far exceed this revenue. Time and capital will make this plant as modern as there is in the land. We want it that way as bad as you. Please feel free to call the office or come by and discuss your telephone needs and problems."

In 1968, death claimed Dr. Linwood H. Denman, 77, one of the principal owners of the Lufkin-Conroe system and president of both the Lufkin and Conroe exchanges throughout most of the 1960s. Dr. Denman, who was a past president of the Texas State Board of Medical Examiners, had followed in the footsteps of his father, Dr. Alexander Madison Denman, as a physician. He was a founding member of the Southwestern Surgical Congress, a member of the state and Angelina County branches of the American Medical Association, and president of the Angelina County Medical Society. He had also served as chairman of the

The Lufkin telephone exchange expanded its Lufkin facilities with a second-story addition in the 1960s. This photograph was made in the 1970's.

Lufkin-Conroe telephone system.

Dr. Denman's death came in late March during a Lufkin banquet at which he was to be the principal speaker. The chief of Memorial Hospital in Lufkin, he was stricken with a heart attack just before he was to address a Doctors Day banquet. Funeral services were held on Tuesday, April 2, at the First Baptist Church in Lufkin with burial in the Denman family plot in Glendale Cemetery.

Marion Clay, right, visits with fellow telephone industry officials at a Chicago convention in 1960.

He was graduated from Lufkin High School in 1907 and attended Washington and Lee University. In 1913 he graduated from Baylor Medical School and entered the practice of medicine in Lufkin. He was a member of the American Medical Association, the Texas Medical Society and the Angelina County Medical Society. He was past president and a current member of the Texas State Board of Medical Examiners. He was a founding member of the Southwestern Surgical College. He was a lifetime member of the American College of Surgeons where he had served as chairman of the Committee on Trauma. He was a former Chief of Staff of Memorial Hospital and was a member of the Doctor's Club of Houston. He was member of the Governor's Advisory Committee. He was a member of Lufkin Lodge #669, Ancient Free and Accepted Masons, the Houston Scottish Rite Cathedral, and the Arabia Temple of the Shrine. He was a member of the Angelina County Chamber of Commerce. He was president and chairman of the Board of the Lufkin Telephone Exchange and the Conroe Telephone Exchange. He was a member of the board of Directors of the First Bank and Trust Company. He was a member of the First Baptist Church.

Dr. Denman was succeeded as president of the Lufkin and Conroe exchange by N.D. Shands, Jr.

Lucy Hendrick recalls Dr. Denman as a "very conservative man who was very careful with the company's money, as well as his own" and Betty Warren, who worked for the company between 1947 and 1992, remembers Dr. Denman as "a man who loved the telephone company more than he did medicine, and treated it like his own baby." He came by the office every day to sign checks, she remembers, and chatted regularly with the employees.

Before the end of the 1960s, Conroe had outgrown its new offices on Phillips Street and acquired an adjacent building owned by *The Conroe Courier*. Some of the telephone company's operations were moved into the building, but Bateman

A Lufkin telephone directory in 1968 featured the new Angelina College campus.

recalled that the building later turned out to be a hazard. "The building starting falling down, and we had to bring in four-by-four timbers to prop up the overhead beams," he said.

Members of the Lufkin and Conroe boards during the 1960s were L.H. Denman, B.H. Denman, Kester W. Denman, Sr., Frank R. Denman, Peyton L. Denman, H.T. Bowers Sr., E.J. Bowers, D.C. Hucherson, N.D. Shands Jr., Kester W. Denman Jr., Shirley Simons Jr., Darrell H. Hamric, and Marion Clay. Clay became president of the two companies in September, 1969.

The end of the 1960s also marked the creation of the Lufkin-Conroe Communication Company, which was created as the parent company of the telephone companies in Lufkin and Conroe. N.D. Shands Jr. was named chairman of the company, Clay was named president, H.T. Bowers was elected vice-president, B.H. Denman was chosen secretary-treasurer, and Peyton L.Denman was named assistant secretary-treasurer. Also named to the board were Frank R. Denman, E.J. Bowers, Shirley Simons, Jr. and Darrell H. Hamric.

George Shotwell, a former saloon manager, was named to manage the Lufkin telephone exchange in 1900.

13 The Shotwell legacy

In 1900, when Dr. Alexander Madison Denman and Judge Edwin James Mantooth created Lufkin's telephone exchange, they turned to George Shotwell, the manager of Chancey's Saloon and Mantooth's son-in-law, to manage the fledgling company.

For the next 51 years, more than half of the company's existence, the Shotwell family remained at the helm, guiding it through the difficult formative years and into the 1950s, when dial systems in Lufkin and Conroe forever changed the way telephone service was delivered in the communities.

One of Judge Mantooth's motivations for starting the telephone company, according to several of his descendants, was to get his son-in-law, George Shotwell, "out of the saloon business and into something respectable."

Whatever lack of experience Shotwell had in the telephone business, he made up in his dedication and hard work. Shotwell was one of only four employees when the telephone company began serving Lufkin and spent long hours—usually from dawn to dusk—stringing lines, installing telephones, servicing the system, handling complaints, and doing everything else.

While Shotwell—a native of Livingston in Polk County—handled most of the physical work involved in running the telephone company from the Denman Building at the corner of North First and Dozier, the financial and legal decisions were managed a block away in the legal offices of his father-in-law, Judge Mantooth, located on the second floor of Lufkin National Bank. The Denman family had invested the authority in Mantooth as early as 1908 when Dr. Denman was killed in an automobile accident.

"Grandpa Mantooth really ran the phone company; my daddy just did what he was told," recalls Shotwell's son, Mantooth (Bug) Shotwell, 83, of Lufkin, who also worked for the company in Lufkin and Conroe. "Grandpa was a real character, and he had a reputation for sternness, but he always looked after his family."

Shotwell said he was given the nickname "Bug" as a child. "I was a skinny little kid with a great big head, and everybody in the family said I looked like a bug, so it stuck," he said.

George Shotwell ran the telephone company almost single-handed until the 1920s when its growth forced him to seek someone with more experience. He turned to B.J. (Ben) Boon, an installer for the Bell telephone system. "Daddy wanted Ben, who was considered one of the best equipment men in the business, but everybody kept telling him Ben wasn't available. But Daddy kept after him and persuaded him to go to work for the Lufkin exchange for $150 a month, which was

Winnie Mantooth Shotwell, telephone manager George Shotwell's wife, was the daughter of exchange co-founder Edwin James Mantooth.

real good money in those days," recalls Mantooth Shotwell.

George Shotwell died in 1923, shortly after persuading Boon to come to Lufkin, but the company kept Shotwell's promise of a job and Boon went to work in the twenties, working for Jean Shotwell, George's son, who became the exchange's second manager.

Mantooth Shotwell was only 10 at the time. "Ben was like a father to me, and he treated me like a son. He was the best telephone man I ever knew, and he did a lot for the exchange," recalls Mantooth. "The hardest job Jean did was to sign his checks; it really bothered him to pay Ben all that money, but he was worth every penny of it."

Boon, said Mantooth, made many improvements that led the way to the company's growth in the twenties and thirties. "He was a behind-the-scenes man and people never knew of his work," said Mantooth.

In 1943, however, Boon suffered a heart attack while fishing on the Neches River. Mantooth found his body. "The two of us had had a little fuss a few days earlier, and it really hurt me that he died before I could apologize to him," he said. "That has bothered me for a long, long time."

Mantooth himself worked for the telephone company between 1932 and 1948, working for his brother Jean. "When I started work, there were only two other people working on the lines and equipment. We did everything from digging holes for telephone poles to putting in switchboards," recalls Mantooth.

Mantooth dropped out of school in the fifth grade, but Judge Mantooth enrolled him in a bookkeeping course at Satterwhite Business School in Lufkin and made him complete the course. "He said it would keep me busy and out of trouble," remembers Mantooth. "When I finished the course, I went to his office, and Grandpa looked at me and said, 'Bug, take a letter for me.' I told him he paid for a bookkeeping course, and I wasn't a secretary."

When Mantooth started work at the telephone company, his first job was to deliver telephone directories in town. A later assignment was to travel eastward on the Angelina and Neches River Railroad to repair telephone lines at Ewing, a sawmill town on the banks of the Angelina River. In the 1940s, when Ewing's sawmill economy collapsed and the town became a ghost, Mantooth helped remove the telephone lines. "I was taking down a line one day, and a stinging lizard dropped into my

Jean Shotwell became manager of the Lufkin telephone exchange in 1923 when his father died.

shirt pocket, and stung the devil out of me," he recalled.

In earlier years, Mantooth's father, George, had hunted birds along the railroad line, carrying his bird dogs with him while he made telephone repairs.

Another assignment given to Mantooth was to repair telephones used by businesses in downtown Lufkin. "We had one old merchant who chewed tobacco a lot and was always spitting into the receiver. I had to wash out the phone and spray it with perfume because it smelled so bad," he recalls.

Working under Boon and Rex Atkinson, another early employee who later became plant superintendent, Mantooth helped install the first switchboard at the Angelina Hotel shortly after the hotel expanded and put telephones in every room. "Rex wanted to work outside and I wanted to work inside, but Jean switched us around, and I wound up working outside and Rex went inside," said Mantooth.

"In those days, we worked from 7 a.m. to 7 p.m., and we'd go anywhere at any time to repair the phone service," he recalls.

When he left the company in 1948, Shotwell turned to a different enterprise, the production of ice. He founded the Shotwell Ice Company in 1962 on a lot across the street from the old Woodland Heights Hospital on Dozier Street (now Frank) and later moved the company to its present site on North Raguet. Shotwell sold the first bagged ice in Lufkin and built a company that became an institution in the city.

Ironically, Shotwell's grandfather, Judge Mantooth, was a major shareholder in another ice company, the Lufkin Ice Company, founded in 1904. An early newspaper article in 1932 described the company in these words: "The Lufkin Ice Company is the most modern in East Texas. There is nothing of this nature that can be compared with it outside of Houston..." Judge Mantooth also had a major interest in another ice company at Livingston in Polk County.

In 1961, the Shotwell era at the Lufkin-Conroe exchange came to an end when general manager Jean Shotwell retired after a heart attack. Betty Warren, a company retiree who worked as Jean Shotwell's secretary in the 1940s, recalls him as "a wonderful man who made all of us feel like a family." Shotwell loved to hunt at Pine Island Hunting Club on the Neches River and often entertained company families with cooked venison and turnip greens. "He loved to cook for us," said Mrs. Warren, "and it was something he did very well."

To succeed Shotwell, the Denman and Mantooth families turned to San Angelo resident Marion Clay, who became the first non-family member to manage the company since its inception.

Born November 6, 1906, in Tom Green County, Texas, Clay came from an old Texas telephone family. His father was in the telephone business for 48 years, so Clay started in the field at a very early age. Accepting the Lufkin Telephone Exchange's offer of employment in 1951, Clay moved his family to Lufkin in April of 1952.

Mantooth (Bug) Shotwell, left, another son of George Shotwell, managed the Conroe telephone exchange in the 1940s. He is shown here with Bob Shaw, owner of the Western Auto Store in Conroe.

Tommy Curry, center, who sold the Montgomery telephone exchange to the Denman and Mantooth families in 1956, poses with Marion Clay, left, and Morris Bateman during a conversion to dial service.

14 Expansions beyond Lufkin and Conroe

While the acquisition of the Conroe Communications Company in 1929 set in motion the catalyst that would provide much of the growth for the Lufkin-Conroe Telephone Exchange, several other smaller acquisitions broadened the exchange's geographic territory.

Three acquisitions, one in Montgomery County and two in Cherokee County, occurred in the 1950's and 1960's.

In Montgomery County, the company purchased the Montgomery telephone exchange from Tommy Curry in 1956. Morris Bateman, who managed the Conroe system before his retirement as executive vice-president, recalls the purchase: "When we acquired the system, there wasn't much equipment that we could use, so we went over there with a construction crew and we spent a year putting in a new system, stringing lines down every road, and going into all of the homes to install new dial phones."

The lines the company acquired, Bateman recalls, were largely unshielded and vulnerable to damages. "During dove season, I had to run up and down the lines between Dobbin and Montgomery to shoo doves off the lines so the bird hunters wouldn't damage the lines," he remembers.

The new dial system was completed on April 19, 1958, with a cutover ceremony attended by Bateman, general manager Marion Clay, installer Jack Mott, Durell Gray of the Conroe office, and other officials.

The *Conroe Courier* reported on the new system: "An Alfred Hitchcock character might dial M for Murder. But Montgomery County residents will dial M for Montgomery from now on. For the first time in its history, the third oldest community in Montgomery County, Montgomery, has a dial telephone system. The new system was initiated at 7 a.m. Saturday after an extensive installation program by the Conroe Telephone Company, which owns the (Montgomery) franchise. Now the 155 phones in the community are on the dial hook-up with the exception of a few phones in the more isolated rural areas. In a matter of days, all of them will be dial operated. Morris Bateman, local manager of the telephone company, said...the new system is operating efficiently. In the past, Montgomery has had its own small switchboard. To reach Montgomery, a Conroe resident had to dial '67' and since there was only one trunk line between the two cities, many times he found the line to be busy..."

Another acquisition in the 1950s resulted in the merger of the Wells telephone system, located in southern Cherokee County, with Lufkin.

In July of 1956, the Wells system was purchased from G.B. Woods, who operated a small switchboard in the back of his grocery and feed store south of Wells. The switchboard had about 100 lines and more than 50 stations serving Wells, Sweet Union, Forestt and Crossroads.

W.E. Bynum was the owner of the Alto telephone exchange before it was sold to the Denman and Mantooth families in 1963.

To run the Wells system, Lufkin officials moved Clifford and Peggy Lowry to the small community. Clifford maintained the lines and made repairs and Peggy operated the switchboard during the day. At night, Clifford slept in the same room as the switchboard with a night bell by his bed to alert him to callers.

"We lived in the back of Mr. Woods' grocery, and rats were always running across the rafters," recalls Lowry. "They were so big they sounded like armadillos."

Both in their twenties at the time, Clifford and Peggy considered the Wells assignment as "a great big adventure." Their child, Connie, turned three while living in the community. "It was a small community, much smaller than it is now, and people were always friendly," recalls Peggy. "I remember one little girl who used to ride a bicycle bare-footed from town to bring me posies." Other Wells residents brought the Lowerys pies, cakes, fresh eggs and milk, and vegetables pulled from their gardens.

Lowry recalls that the telephone lines in Wells "roared a lot, and I kept wondering why until one day I ran across a line connected to a barbed wire fence." To avoid stringing an expensive new line, someone had linked the line with the wire fence to complete the circuit. "It seemed odd, but it worked," said Lowry.

Some of the lines in the community were strung so low that log trucks with a high stack of logs would tear down the lines, said Lowry. "We had one logger who kept pulling down the lines, and I had to call the law on him, and he apologized to me. He apparently didn't know he was doing it," said Lowry.

The Lowrys stayed in Wells about a year or until the Wells system was linked with Lufkin and the old switchboard was closed down. Clifford, who had joined the company as a full-time employee in 1949, retired in 1986 as the head of the company's PBX sales group.

Following the connection of the Wells system with Lufkin, the Wells City Council on September 29, 1958, gave the Lufkin Telephone Exchange the authority "to construct, reconstruct, repair and maintain a telephone exchange and all necessary and feasible appurtenances" in exchange for 2% of the gross receipts received from subscribers living in the Wells city limits. Telephone rates in the community were established at $7.80 a month for a one-

W.E. Bynum, right, owner of the Alto telephone exchange, supervises the work of operators at the Alto office in the 1930s.

party business telephone, $5.40 for a one-party residence telephone, $4.50 for a two-party phone, $3.75 for a four-party line, $1.75 for each business telephone extension and $1.25 for each residential extension.

In 1963, the Lufkin exchange purchased the Alto telephone exchange from W.E. Bynum, a former Lufkin telephone company employee who had bought the Alto system in 1920 from C. Hill Fitts. The latter founded the exchange in 1916 by selling enough capital stock to his friends to build a small magneto crank telephone system serving less than 100 subscribers. Bynum was one of the original stock purchasers.

In 1920, when he purchased all of the stock in the Alto company, Bynum loaded his family and belongings into a railroad boxcar for the trip to Alto. He often told relatives: "Our milk cow was in one end of the boxcar and the family and all of our furniture were in the other end."

Bynum was the son of Lufkin pioneer W.D. "Billy" Bynum, for whom Bynum Street is named. Bynum lived on the street in a large, unpainted house and owned a cornfield where Lufkin's Memorial Health System complex stands today. Bynum, known as "Grandpa Billy," came to Texas with a wagon train from Alabama, served as an Angelina County commissioner in the early 1900s, and is remembered by his relatives as a talented violinist who often played on the street corner near his home.

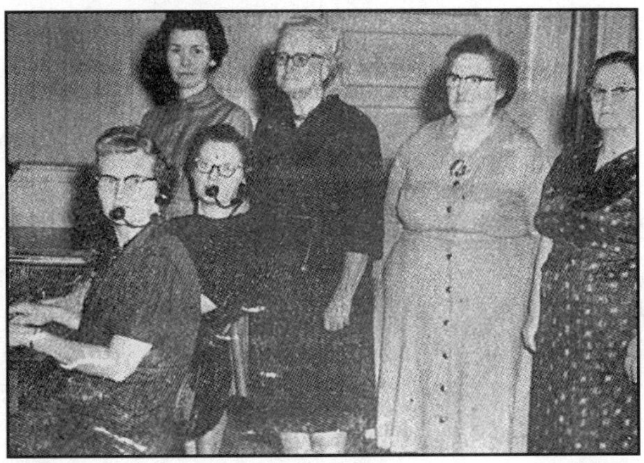

The office staff and operators of the Alto telephone exchange.

W.E. Bynum operated the Alto system with the help of his son Edward and Pauline Bynum, the wife of another son, Marshall, an Alto electrician. When Lufkin acquired the Alto system in 1963, Edward and Pauline remained as employees at Alto, both retiring in 1983.

W.E. Bynum operated the magneto crank system at Alto until the 1950's when he converted the exchange to a common battery operation which allowed subscribers to pick up their receivers and make connections with the assistance of switchboard operators.

The central switchboard was located on the second floor of a building housing a grocery store at the corner of U.S. Highway 69 and Texas Highway 21 in Alto. Edward Bynum recalls hauling loads of wood up the stairs to the telephone office during the winters to keep a fire going in a wood stove. The office also housed a bed where the night operator slept until she was awakened by a bell positioned near the bed.

"In those days, before we had answering machines and call forwarding, the operators did everything for the customers," recalls Pauline Bynum, who worked as an Alto operator in the early 1960's. "If you had to be away from home, you could call the operator and tell her you would be shopping, visiting a neighbor, or even taking a bath, and she would hold your messages until you were able to answer the phone."

Alto's operators were also in charge of the town's fire alarm. If a fire broke out, they pulled a string connected to a fire whistle mounted atop the telephone offices. "After a fire, our switchboard would light up like a Christmas tree; everybody wanted to know where the fire was," said Mrs. Bynum.

Long distance calls from Alto were passed by operators to Southwestern Bell in Jacksonville. "Mr. Bynum never made any money on long distance. In fact, he usually lost money," recalls Pauline.

The month after Lufkin acquired the Alto system, customers found themselves with two bills. "Mr. Bynum had been billing people at the end of the month, but Lufkin started billing people in advance of each month," recalls his daughter-in-law. "We had a lot of unhappy customers in Alto before we got things straightened out."

In 1965, about 18 months after it acquired the Alto exchange, Lufkin completed a new direct dial telephone system for the community. W.E. Bynum, who was in failing health and died later that year, was given the privilege of making the first dial call to Lufkin in January, 1965. To house the new dial system, a new building was built beside the Alto Post Office with a drive-in window and two pay phones. Alto also became the first city in Cherokee County to have a direct dial long distance system, which enabled Alto residents to dial Lufkin directly without going through an operator. Alto to Lufkin calls remained toll calls until about 1995.

Lufkin financed the Alto dial system by borrowing $220,000 from First State Bank and Trust of Lufkin. Marion Clay wrote this note and inserted it into the company files: "Delivered certificate No. 1 for 3,000 shares of Alto Telephone Company to Ned Shands, Jr. (the company's attorney) to be delivered to Stromberg-Carlson Company as part of mortgage on the Lufkin Telephone, Inc., loan. A signed receipt will be returned from them."

Pauline Bynum, who remained in Alto after her husband Marshall died in 1991, has collected many mementos from Alto's telephone history, including the stock shares from 1916 and several early telephone directories, including one from 1949 which includes the names of many pioneer families and businesses.

The Alto exchange was operated as a separate telephone company until 1986 when it was merged with the Lufkin and Conroe exchanges to become Lufkin-Conroe Telephone Exchange, Inc. Serving as directors of the Alto exchange during its ownership by the Denman and Mantooth families were E.J. Bowers, Marion Clay, Peyton L. Denman, H.T. Bowers, Sr., B.H. Denman, James E. Newman, G.I. Ross, N.D.Shands, Jr., Kester W. Denman, Jr., Edwin M. Simons, and Linwood D. Newman. Serving as presidents were Clay and Ross while Kester W. Denman, Jr., was chairman.

This frame building housed the Montgomery telephone system when it was converted to dial service in 1958. Standing on the doorsteps are, left to right, Durell Gray, Louis Calfee, and Avon Knighton. The man on the mule is unidentified.

At Conroe, the telephone system was expanded to the Riverbrook community on August 26, 1970, to Grangerland on September 25, 1971, and to Cut & Shoot community on March 19, 1972.

In announcing the expansion plans for the 1970s, Conroe manager Morris Bateman told *The Conroe Courier*, "We believe the best telephone service is due Conroe and the surrounding area, and we are striving to provide it."

Bateman recalls that the company barely missed an opportunity to become the telephone provider for all The Woodlands, a rapidly-growing community south of Conroe with a population in excess of 100,000 in the 1990s.

"Before Houston started growing up Interstate 45, the area between Spring and Conroe was mostly rural countryside. An old man over in Tomball, who owned the telephone company there, had strung a line down the railroad track from Tomball to serve a pay station in Spring. In those days, if you had a single telephone in a community, you had the right to serve the area," said Bateman.

"The old man had to service the line to Spring, and one day he came over to Conroe, told me he was tired of fooling with the line, and would sell us the line, poles and eight telephones for $500," said Bateman. "But money was short in those days, and we were having trouble taking care of our subscribers in Conroe, so I turned him down."

If Bateman had bought the line, the Conroe company would have had the telephone service rights from Conroe to Spring. Today, that area—which is serviced by Southwestern Bell—includes the southern part of The Woodlands and dozens of other rapidly-growing bedroom communities scattered along Interstate 45. The Conroe exchange today serves the northern one-third of The Woodlands, the portion inside Montgomery County.

Durell Gray, a lineman at Conroe in the l950s.

15 Tales from the field

Underground cables and fiber optics have diminished the need for linemen in today's modern telephone company, but in the early years of the Lufkin-Conroe system they were the backbone of a network that depended almost exclusively on poles and overhead wires.

Between 1900 and the 1950's, when dial telephone service came to Lufkin and Conroe, subscribers not only knew their central operator, but usually came to know their lineman and repairman as well.

"You came to know a lot of people living in the country," said W.C. Tipton of Apple Springs, a lineman who worked for the Lufkin telephone exchange between 1952 and 1992. "People were always interested in what we were doing up in the air."

In the early days, Tipton and other linemen worked primarily with lead-sheathed cables. The lead covered paper-insulated copper wires which transmitted telephone conversations.

"We had a lot of problems with the old lead cables," recalled Tipton. "For some reason, squirrels would eat the lead, probably because of the salt, which opened

Telephone repairmen pose for a photograph. Standing, left to right, are Joe Wadsworth, Obie Read, Marvin Schmidt, Kenneth Calhoun, W.C. Tipton, David Colwell, Norman Tindall, Melvin Haak, Dalton Conner, Robert Beavers, Gerald Glover, Ike Smith and Robert Brock. Kneeling, left to right, are Joe Noel, Melvin Burrous, Johnnye Barrett, Jack Mott, Monroe Ivy, Bobby Coleman, Clifford Lowry, Johnnie Weatherford, Walter Lawrence, and John Harris.

the paper to moisture and caused shorts in the wires. We spent a lot of time dealing with squirrel damages."

Another retiree, James T. (Jack) Mott, says squirrels were so bothersome in one area of Lufkin that linemen started trapping squirrels in a city park. "One day, a woman came out of her house, crying that someone had shot her pet squirrel," said Mott. "That ended the trapping program."

Workmen check telephone equipment in Lufkin. Left to right are Bill Riley, Clifford Lowry, Jack Mott, Lewis Calfee, Norman Tindall, and Maxie Pearson.

Lightning strikes and contacts with electrical wires were also troublesome for linemen. But water was the biggest enemy, said Tipton. "If a 90-volt ringing current line came out of an air conditioned building and into the hot air, you could count on moisture being in the line," he said. "This was a particular problem with the old Denman Building, where the telephone company was first located. We spent a lot of time there fighting moisture."

Splicing a cable to repair damage was time-consuming and often took the best part of a lineman's day.

"First, we had to chip away the lead on the outside of the cable, remove the paper covering, and strip and twist the wire. Then, we used hot paraffin, which was sent up to the line by a man on the ground, to draw out the moisture in the cable. Then we had to put a muslin wrap around the wire, put a lead sheathing back over the splice. Finally, we had to fill the line with nitrogen dry gas to see if there were any bubbles left in it," said Tipton.

The cables were operated under pressure and alarms were installed to alert telephone company repairmen to malfunctioning lines. "When the underground cables came in the 1970s, a lot of the work we did as linemen was eliminated because most of the service interruptions were done away with," said Tipton.

During Tipton's days, linemen erected their own telephone poles and strung their own wires. "It usually took three or four men to carry a 25 or 30-foot wooden telephone pole to the site. We'd then lay down the pole with one end resting against the hole, and use pipe poles to push the pole in the air until it fell into the hole," he said.

A lineman's work was often hazardous. "On one job at the paper mill in Lufkin, Joe Wadsworth and I were taking down an old wire inside the mill and winding it onto a reel. The wire got mixed up with a live electrical line, and slung melted rubber all over our faces. We thought we were dead," recalls Tipton.

On another occasion, Tipton was working on a telephone pole near Fairview. A property owner had apparently burned trash against the base of the pole, leaving it damaged. When Tipton disconnected the wire, the pole began falling and almost threw Tipton to the ground.

Linemen in Lufkin and Conroe, and the communities surrounding the two cities, worked in all types of weather. "The cold weather was the worst," remembered Tipton. "When you went up a pole during the winter, the air was colder the higher you went, but we had to stay with the job until it was done. If rains were falling, you usually had to break the ice on your slicker suit before you came down."

Company retiree Betty Warren remembers when lineman Johnny Weatherford came to her office with a $12 invoice from the Western Auto store for a new tricycle. Puzzled, she asked Weatherford why he had submitted the bill.

"Well," he said, "it's like this. We were out digging a ditch, and some kids playing on their bikes and tricycles were watching us, and I backed over one of the kids' tricycles with my truck. I figured I was lucky to pay for just a tricycle, and not a kid."

When the Lufkin telephone exchange converted to a dial system in the early 1950s and built a new exchange building to handle the system, Tipton tied in all of the cable to the new building, and Mott did much of the testing for the new equipment.

A rural telephone used in the 1950s.

Mott started his career as an installer's helper, working with linemen, repairmen, and other field workers out of an old building on what the work crews called "One Thousand Alley," off Angelina Street. "At that time, in the forties, we had nine men, one old flat-bed truck, a jeep, and a pickup with an installer's box mounted on the back. Inside the building, there was an old pot-bellied oil stove that kept us warm in the winter," he said.

In the years during World War II, new equipment was scarce. "At Keltys, we grounded the telephone line with the earth to complete the circuit, but when the earth got dry, the circuit wouldn't work very well, so some of the guys went out to Keltys and poured water on the ground where the line connected with it. There was a lot of roaring in the line, but it worked," said Mott.

During the war years, the company had 12 to 15 households on some lines. "Everything worked with a series of rings, such as shorts and longs, and every customer had his or her own ring," said Mott.

Mott worked in a series of jobs before he retired in December of 1987. In addition to his first job as an installer's helper, he was an installer, wire chief, switchman, plant superintendent, and customer service manager. "I enjoyed every job I had, and I took a great deal of pride in all of them," he said.

Mott remembers a dial relay that became a problem every place it was used. "I took it out of Diboll when it didn't work right, we tried it in Wells when we bought the telephone company there, and later moved it to Apple Springs. It never seemed to work right anywhere," he said.

Veteran telephone installer Walter Lawrence, who worked for the Lufkin exchange from 1952 to 1994, remembered the dial conversion vividly. "One of my first jobs in 1953 was to install dials on the subscribers' phones. Norman Tindall and I had a suitcase full of dials, and we walked down the streets of Lufkin, installing dials on the phones. Many of the customers didn't understand the dials, so we had to explain how to use them."

A lineman performs a balancing act during a repair job.

Lawrence, who says he "enjoyed the fire out of working for the telephone company," remembers that people were more trusting in the 1940s and 1950s, when he worked as an installer and repairman. He spent 14 years of his service in Diboll.

"In those days, people would call us about a phone that wasn't working. Then, they'd leave their doors open for us to come and make the repairs at our convenience. Nobody locked their doors then, and nothing was even stolen," he recalls.

On one occasion, a Diboll resident and friend of Lawrence reported a malfunctioning telephone. Lawrence went to the home and found a meal of T-bone steak and mashed potatoes sitting on the dining table. "I fixed the phone and sat down and ate the meal," said Lawrence, "but I left him a note thanking him for the supper."

Lawrence had a gift for remembering the telephone number of virtually every subscriber in Diboll and other communities where he worked. "I eventually lost the gift as I got older, but in those days I could match up a customer with his telephone number and tell you where the customer lived, too."

While eating at Diboll's old Pine Bough Restaurant, Lawrence overheard a customer ask the restaurant manager if she knew the telephone number of a Diboll resident. Lawrence spoke up, easily rattling off the number and address.

Dogs were a constant source of irritation for Lawrence and other repairmen. "One day, I went to the home of a customer who had a black and tan dog. The dog didn't bother me when I went into the house, and didn't bother me when I made two or three trips back and forth to my truck for tools. But when I finished repairing the phone, and walked back to the truck to leave, the dog nailed me, ripping my trousers and taking a good bite out of my leg," he said. "I went to the Pine Bough, where the customer worked, and told her. A couple of days later, she bought me a new pair of pants."

On another repair trip in the Beulah community, a customer had a dog which started snapping at Lawrence. "I picked up a tree limb and gave him a good whipping, and he never bothered me again," he recalled.

Lawrence remembers another customer who kept calling him and other repairmen, complaining that her telephone had been wire-tapped by spies and wanted it checked.

When he went to work as an installer, Lawrence not only made repairs, but was required to collect for the work as well. "We turned our money in every Friday, and one week I came up two dollars short. The company wanted to take it out of my pay, but I only made 75 cents an hour, so I went over my route for the week, and found a lady I had forgot to charge. My boss told me it was a good lesson for me; that I wouldn't forget the next time, and I didn't," he said.

In the 1940's, company installers, linemen and other men who worked in the field had to purchase their own tools and spent much of their time repairing trucks and other equipment they used to make their rounds. "If you had a flat, it took an act of Congress to buy a new tire. One day I spent six hours in Sweet Union, waiting for someone to bring me a spare tire when I had a flat," said Lawrence.

Things changed with the arrival of new executive vice-president G.I. (Larry) Ross in the 1970s. "The first thing Mr. Ross did was to make sure we had the right equipment to do our job, and we never again had to buy our own tools," said Lawrence.

When Morris Bateman, who preceeded Jack Mott as wire chief at Lufkin, came to work in Conroe in 1948, the system had only one other male employee, Bob Ellisor. "The company didn't have a single truck of its own, but I bought a truck myself, and Jean Shotwell (the manager of the Lufkin and Conroe exchanges) agreed to pay me $100 a month for the use of my truck," he remembered.

"But eventually we got so busy that one truck wasn't enough, so we started paying Bob Ellisor for the use of his truck, too. We didn't have any construction crews, so we had to do everything ourselves."

One of Bateman's first jobs was to expand the number of rural lines served by the exchange. "I rented a jeep with an auger on the back, which was used to dig

Telephone company personnel during a union meeting in Livingston. Standing, left to right, are Jack Mott, Clifford Lowry, A.S. Carrington, Drayton Brown, Wallace Nutt, Robert Brock, Obie Read, unidentified, Joe Wadsworth, Maxie Pearson, Ike Smith, and Johnnie Weatherford. Seated, left to right, are a lady named Helen (last name unknown), Matt Hand, four unidentified employees, Louis Calfee, George Conner, and Tom Dominey.

fencepost holes, and we'd take the jeep out in the country to dig pole holes for the telephone lines. But the thing wouldn't go down in the earth very good. I had a big guy working for me, and he'd hop on top of the digger, and it would go so far down that it would stick, and we'd have to dig it out," said Bateman.

Each summer, the Lufkin exchange sent a work crew from Lufkin to Conroe to help Bateman build at least one new line to serve new customers. Bateman started using the ground in rural areas to serve as the grounding wire instead of running a second wire. "But when the REA (Rural Electrification Administration) did the same thing, our telephone lines roared like crazy, and we had to string new wires to stop the noise," he said.

Near the end of the 1950s, the Lufkin exchange bid farewell to a legendary employee, 50-year lineman Matt Hand, who had worked on virtually every line in the Lufkin area since joining the company shortly after the company was founded.

Clifford Lowry, who worked with Hand, penned a poem to Hand's service. It has been passed around for years within the company as a tribute to a telephone lineman:

ONE MAN CREW

In the days when telephony was young,
There was a man who's feats went unsung.
Now comes the time when all shall hear,
Of this man's deeds, who worked without fear.

Up in the morning at the crack of dawn,
He'd weed his garden and mow the lawn.
Walking to work was a joy to him,
Chatting with friends while the sun was still dim.

Shouldering a pole, and a cross arm or two,
You could see him walking, till he passed out of view,
Down the tracks for ten miles and a quarter,
Resting only when he met the farmer's daughter.

This may sound funny, and a little erratic,
But parts of him were really automatic,
After setting the pole, he tamped till the earth shook,
Then sailed up the pole with a knee cap and one hook.

He could hang cable like a regular fiend,
Then wrap it up with some rusty fourteen,
When he cleared a right of way the trees fairly flew,
To us this seems strange, for he was a one-man crew.

He has made more friends than anyone around,
For he made every coffee shop in town.
All jokes aside, he was a man strong, and fine.
He could repair phones and string number nine.

Are you still wondering about his identity?
Who strung wire from here to eternity?
Well, take the load off and set in the sand,
For it was none other than Matt L. Hand.

Kester W. Denman, Sr., was an astute lawyer and businessman, but he was also a humorist and storyteller.

16 Chitlins and catfish chewers

Kester W. Denman, Sr., the son of telephone pioneer Alexander Madison Denman, served his father's telephone company as an attorney, president during a brief period in 1944, and as a member of the board of directors.

He was an astute lawyer and a serious businessman with a focus on the family's principal business enterprise.

But family members also knew him as a humorist and storyteller, as evidenced by an article he wrote for the *Lufkin Daily News*, apparently in the 1940s. The article was titled, "Chitlin and Catfish Chewers Meet, Eat, Organize and Elect To Perpetuate Jonah-Founded Order."

The article follows:

The club mania, which has in such virulent form and with such sinister motives attacked the American People within the last decade, reached its apogee last Sunday night at the Ruby Cafe in the organization of the Magnum Intestinum Chapter of the Old and Odiferous Order of Chitlin and Catfish Chewers.

Those present and taking part in the organization were: R.M. Feagin, Emma Feagin, Jean Shotwell, Garvis Shotwell, L.H. Denman, Ruth Denman, K.W. Denman, Cleo Denman, J.E. Walker, Devilla Walker, and Mrs. Winnie Shotwell. Ned Shotwell, Buddy Shotwell, Peyton Denman, Caledonia Denman and Mary Jean Shotwell were also present.

The menu consisted of chitlins, crackling bread, pickles, celery, crackers and coffee.

After finishing the meal, the group went into permanent organization with the following as its officers: Jean Shotwell, Exhausted Caecum: R.M. Feagin, Impotent Duodenum; and Mrs. Winnie Shotwell, Juvenescent Jejuno-Illeum. In other groups these officers correspond to president, vice-president, and secretary and treasurer, respectively.

It seems that this organization is one of the oldest societies in the country, having been organized at Walla-Walla, just prior to the Civil War by a swine-herd who claimed to be a direct descendant of the Prodigal Son.

There are many chapters throughout the United States, and the organization here was fostered by Mrs. Winnie Shotwell who had long been an advocate of the efficacy of chitlins and catfish as the substitute for the animating proclivities of medicines and certain advertised products.

Mrs. Shotwell explained the origin of the organization, stating that it was originated by Jonah during the period of his sojourn in the belly of the fish; and that during the time the Israelites were captives in Egypt, being unable to obtain the roe of the fish (of which Jonah had become enamored), they were compelled to resort to the use of the chitlin as the principal dish at their banquets, which were, of course, held sub rosa.

Kester W. Denman, Sr., conducts a patriotic ceremony at the Angelina County courthouse in Lufkin in the 1950s.

The liberals among them applying the General Welfare Clause to the Law of Moses, determined that the chitlin was not a part of the flesh of the swine.

Mrs. Shotwell further stated that she was really more interested in the catfish branch of the organization than in the chitlins due to the following part of a verse of Scripture which she ran across on Sunday afternoon while reading her Bible. "Go thou to the sea, and cast a hook, and take up the fish that first cometh up; and when thou hast opened its mouth, thou shall find a piece of money."

R.M. Feagin gave an interesting talk on the best method of cooking a drum fish. Dr. L.H. Denman was called upon for a talk in respect to the part that chitlins played in the progress of civilization.

Although his tones, due to his chitlin content at the time, were "guttural," he evidenced a rare intestinal fortitude and an intimate knowledge of the subject. The doctor stated that the spinach eaten by Popeye is nothing more than metamorphosed chitlins; that the chitlin was the only sure medical remedy for Hepaticocholecystostcholescystenterostomy; and that the vitamin E content of the chitlin is not exceeded by any other known food or medicinal preparation.

Several Lufkin citizens who had indicated their desire to become members were not present. Wood Newsom was again called to Liberty. Mrs. Edna Winston, who is the "little Edna Winston," who celebrated her fourth birthday in 1904, had a previous engagement. Her husband, W.D. (Mute) Winston's excuse was that it was his night off and he could not be bothered by an inanimate chitlin. Helen Denman was compelled to attend a meeting of the Sheltered-From-the-Glare-of-Light Story Tellers Club. Joe C. Denman accompanied her with his flashlight.

It is planned by the group to have two meetings each year—one in the fall at which chitlins will be the main diet and one in the spring when catfish will make up the item on the menu. The membership is hopeful that there will be many additions during the ensuing year and that the club will be the foremost agency in promoting the growth of Lufkin and strengthening the morale of the people.

Kester W. Denman, Sr., passed away in 1966 at the age of 78. He was a retired attorney and was employed by the Lufkin National Bank for many years. He was a graduate of Washington and Lee University with a B. A., and the University of Texas with a L.L.B., he was Chairman of the Board of Directors for the Salvation Army for many years, member of the Texas Bar Association, American Bar Association and Angelina County Bar Assoc., Charter Member and First President of the Lufkin Noon Lions Club, member of the Board of Directors of Lufkin National Bank, and Lufkin Telephone Exchange and Conroe Telephone Co. He was an Elder in the First Christian Church.

Lufkin telephone worker Chuck Lundstedt inspects some of the exchange's new equipment in the mid-1970s.

17 The seventies: "The challenge of the job intrigued me."

Just as Marion Clay's leadership had helped the Lufkin-Conroe Telephone Exchange make the transition from the operator era to dial service, the arrival of G.I. (Larry) Ross in 1970 began to shape another significant era—one characterized by new technological advances and growth in its two core communities, Lufkin and Conroe.

Born in Coryell County on a ranch that has been in his family since 1824, Ross graduated with a BBA degree in business in 1957 from Howard Payne University in Brownwood, followed by graduate work at Michigan State University in engineering and accounting and additional graduate work in finance at Southern Methodist University. He received his MBA in finance in the early 1970s.

Ross began his career in 1956 as an engineering assistant for Southwestern States Telephone Company in Brownwood, remaining there until 1965 when the company was merged with General Telephone Company. Ross then moved to Dallas, joining the Bell Telephone system as a senior engineer. He stayed there two years and moved to Continental Telephone Company, also in Dallas, where he was elected vice-president of finance for the company's Southwest region consisting of Texas, Arkansas, Oklahoma, Louisiana and New Mexico.

Larry Ross joined the Lufkin-Conroe telephone system as executive vice-president and was later promoted to president and chief executive officer.

In February of 1970, Ross received a telephone call from Kester W. Denman, Jr., a director of the Lufkin Telephone Exchange and the grandson of both Dr. Alexander Madison Denman, and Judge Edwin James Mantooth, who founded the company in 1898. Denman told him Marion Clay was retiring and asked if Ross would be interested in succeeding him.

"I told Mr. Denman I had a good job with a promising future, and wasn't interested," remembers Ross. "But three months later, he called back and asked me to meet him on the 11th floor of the Bell Telephone Company building, where he was an attorney for Bell. He wanted me to accompany him to Conroe, where the company was having a lot of problems serving the community," said Ross.

Larry Ross, right, shows Lufkin mayor Pitser Garrison some of the new digital switching equipment installed by the Lufkin telephone exchange.

"I flew into the Houston airport, met him, and we went to Conroe, I looked over the exchange, and concluded it was a network that needed a lot of work. It was deteriorating and clearly inadequate to serve the town. It was the toughest kind of situation. The town was growing fast and the company wasn't able to modernize its facilities to keep up with the growth," said Ross.

"On the way back to the airport, I told Mr. Denman I didn't want the job," recalls Ross. "But Mr. Denman looked at me and said, 'What's the matter. Don't you think you can handle it?' The challenge of the job intrigued me and, there at the airport, we negotiated a deal."

The enormity of the challenge struck Ross when he arrived with his family in Lufkin on September 1, 1970, to become executive vice-president of Lufkin-Conroe's four operating entities, Lufkin Telephone Exchange, Conroe Telephone Company, Alto Telephone Company, and Lufkin-Conroe Communications Co. Clay remained as president.

"The only thing the company had going for it was two fairly safe city franchises in Lufkin and Conroe," he said. "The Lufkin telephone exchange was sub-standard and Conroe's network was very poor. There was a complete absence of a budget and financial plan. The company didn't have an engineering department, service was poor, and there wasn't much of an organizational structure. There were a lot of good people working for the company, but they needed training."

The 1970 Lufkin telephone directory featured the Lufkin zoo and its miniature railroad.

In the following months, Ross instituted a budget and finance plan, hired an engineering staff, put new supervisory personnel in place, and began shaping a plan for modernizing the Conroe and Lufkin exchanges.

The changes he instituted were unlike anything seen by the Denman and Mantooth families. Since World War II, the company had been managed largely by Dr. L.H. Denman, a son of Dr. Alexander Madison Denman, one of the founders. Dr. Denman owned 20 percent of the company's stock and "ran it out of his hip pocket," according to a company retiree.

But Ross had the unanimous support of the two families for his plan to modernize the two exchanges. "It was necessary to borrow ten million dollars to make the improvements that were needed, and that was a lot of money for the seventies, but the board wasn't reluctant to make the commitment," he said. "We signed a 25-year note with Hancock Insurance Company at an eight percent interest rate, which was a big leap for the company."

In September, 1970, the Board of Directors for Lufkin-Conroe Communications Co., the parent company of the Mantooth and Denman telephone interests, consisted of Darrell Hamric of Dallas, who chaired the board; Shirley Simons, Jr. of Tyler, Harry Bowers, Sr. and his son, E.J. Bowers, Dr. Byford Denman of Lufkin, Kester W. Denman, Jr., Dr. Peyton L. Denman of Lufkin, Dr. Frank Denman of Houston, Marion Clay and N.D. Shands, Jr., vice-president and general counsel.

One of Ross' first efforts was to deal with the strain put on the system by the rapid growth of Conroe and Montgomery County, as reflected in a gain of 8,768 subscribers in 10 years, giving the company 10,896 subscribers in the Conroe area.

"I had been on the job only a few months when the mayor of Conroe called a public meeting to talk about the town's telephone service," remembers Ross. "All I could do was tell the public the truth. The service was poor and we needed a whole new network, but we were starting and would have things turned around in three years."

In a notice to customers in the spring of 1970, Ross promised Conroe subscribers improvements in their services. A notice to subscribers said: "The conversion will be done in two phases. First, the 539 numbers will be converted to new equipment...followed by the 756 numbers converted...particularly noticed during this conversion will be a momentary silence after dialing, which will exist until all equipment is changed."

In August of 1970, the company expanded its telephone service to the Riverbrook community, serving 568 subscribers. "Riverbrook was our first big satellite office, and it greatly improved our service to River Plantation, which was the first big subdivision built south of Conroe. This was before Lake Conroe was built, and before the big subdivisions were built around the lake," said retired Conroe executive vice-president Morris Bateman.

The improvements made it possible for the company to keep pace with the 150 to 200 stations being added a month in the Conroe area, largely the result of River Plantation's growth.

But the community still had dissidents to deal with. In the spring 1971, Conroe City Council meeting, irate telephone customers protested a proposed increase in telephone rates—a continuation of customer problems the exchange had endured for years in Conroe.

On September 25, 1971, the company added Grangerland to the system with a new service center, serving 584 customers.

A month later, Conroe Mayor R.A. (Mickey) Deison, who had been feuding with the telephone company over customer service and a proposed rate increase, dropped a bombshell in the laps of Ross and others. Deison proposed making common stock in the phone company available to Conroe investors, moving the company's corporate offices from Lufkin to Conroe, and providing extended area service to Houston. Ross and Bateman were present during Deison's announcement and, rather than commenting on the proposals, left the meeting.

While Deison's proposals were endorsed by his fellow City Council members, the proposals, for the most part, were never implemented.

In 1972, Ross moved to soften customer complaints in Conroe by announcing an improvement program amounting to $1,650,000 for the Conroe exchange, including the construction of new buildings, equipment for a central exchange, outside plant improvements, new vehicles, and new work equipment. "The 1972 budget is a continuation of many of the programs in progress to give our customers better service," said Ross. "It is the plan of the Conroe Telephone Company to continue programs of expansion and upgrading until service is at maximum efficiency."

The occasion of the 1000th electronic PBX system installed by the Lufkin telephone exchange was marked by this photo at Lufkin National Bank in the l970s. Left to right are bank president Pitser Garrison, Tom Worthy and G.I. (Larry) Ross of the telephone company, and Robert Richardson of the bank. The individual at the desk, a secretary at the bank, is unidentified.

On March 19, 1972, the Cut 'N Shoot community was added with 457 customers, the Montgomery exchange was moved into a new building, and Lake Conroe was cut into service with 349 customers by March 4, 1973. By the end of the decade, the company had satellite offices in Riverbrook, Grangerland, Cut 'N Shoot, Lake Conroe and Walden. "These offices enabled us to serve our customers from these offices, rather than run lines everywhere from the office in Conroe," said Bateman.

In 1973, Ross also announced that a progress report on all improvement work by the Conroe exchange would be made at quarterly meetings between telephone company officials and members of the Montgomery County Chamber of Commerce. Ross said the growth of Montgomery County had kept telephone company officials "working around the clock" to give subscribers adequate service and to extend new service as quickly as possible. Ross also told the community that expanded metro service dialing, a service sought by some community groups, would penalize more than 85 percent of Conroe's subscribers to benefit a small minority of businessmen who wanted toll-free calling to Houston. Mayor Deison had been a proponent of the service.

In a Conroe speech, Don Buckalew, president of the Chamber of Commerce, said his organization had two high priorities, the south loop and telephone service in the county. "These two items are the main topics of nearly all the calls of complaints coming into the chamber's office," he said.

Ross said he was well aware of the problems caused by the growth of Montgomery County. "If we could hire the technicians we would like to hire, things would be a little better, but craftsmen in this field are hard to find and very few want to move to a metropolitan area where growth is as great as it is here in Montgomery County," he said.

Meanwhile, in Lufkin, the company was emerging from a bruising battle with the City of Lufkin over rate increases.

The battle had begun in September of 1969, before Ross' arrival, when Marion Clay asked the City Commission for an increase that would give Lufkin the highest rates of any East Texas city. However, the company was offering more services, such as direct distance dialing, than many of the cities.

In January of 1970, the City Commission denied the increase, but Clay and his attorney came back with another rate increase proposal in October. A newspaper headline reported that the city was "cool" to the new request and Mayor Pitser Garrison observed that "the negotiations (between the city and the telephone company) will be rather lengthy."

The same newspaper also carried the news that telephone workers in Lufkin had returned to work after a walkout following a dispute over the right to wear a union button. The walkout involved about 90 percent of the 141 workers in the company's bargaining unit in Lufkin, Conroe and Alto.

The rate dispute continued until July of 1971 when the telephone company obtained an injunction in 159th District Court at Lufkin to institute increased rates in Lufkin. The court decision cleared the way for the company to impose higher rates after twice being refused increases by the city. The increases, the first for the company since 1957, were necessary to provide for expansion of the company's facilities in Lufkin, Ross told *The Lufkin Daily News*. He pointed out that the company had already spent $750,000 of a $1.2 million construction and expansion budget.

But in September of 1971, the City Commission repealed the ordinances dealing with telephone rates, adding a new wrinkle to the fight, and in February of 1972 proposed that the city buy the Lufkin telephone franchise when it expired in 1973. Company president Marion Clay, however, said the company wasn't for sale.

In March of 1972, the City Commission finally passed an ordinance setting new rates for the telephone company. Ross, however, said the rates were not fair and violated the court injunction of 1971. The city disagreed, saying the ordinances ruled invalid by the court had been repealed.

In 1974, the Conroe and Lufkin exchanges, as well as others in Texas, watched with interest as the Texas Supreme Court handed down a decision upholding the right of Texas cities to regulate extended area telephone service.

The decision upheld a verdict initially handed down by District Judge Pat Boone in a lawsuit filed by the city governments of Littlefield, Amherst, and Anton. The three cities had sought to block a rate increase amounting to as much as 60 percent by General Telephone Company for extended area service provided in the three towns. Extended area service provides local service, plus toll-free calls, to certain towns for a fixed monthly fee.

Ross moved from executive vice-president to president of Lufkin Telephone Exchange and Conroe Telephone Company in May of 1973. He was also promoted to president of Lufkin-Conroe Communications Co., the parent company of the Lufkin and Conroe exchanges, in March of 1974.

A year later, in 1975, death came to Ned D. Shands, Jr., 60, a member of the board and legal counsel for the Lufkin Telephone Exchange and a member of the Board of Directors for the First Bank and Trust Co. He was president of the Angelina and Nacogdoches Counties Water Control District Number One.

He was a practicing attorney and past president of the Angelina County Bar Association. He was a member of the Texas and the American Bar Associations. He was also a former Lufkin City Attorney and had been admitted to practice before the United States Supreme Court.

He was past president of the Lufkin Noon Lions Club and past president of the Lufkin Board of Education. He was also currently serving as chairman of the Long Range Planning Committee of the First Baptist Church.

Larry Ross, right, poses for a photo with H.J. (Bubba) Shands, a Lufkin banker who was instrumental in financing LCTX improvements in the 1970s.

In 1975, Montgomery County officials made public a prediction that the county's population would increase by 70,000 in the coming decade. Dennis Berryhill, a grants administrator for the county, compared the county's growth along Interstate 45 between Conroe and Houston to similar development between Houston and Galveston. Don Moore, manager of the Chamber of Commerce, was more optimistic than Berryhill. He said his figures showed the county would have 250,000 residents, compared to 100,000 in 1975, at the end of 1985. "The nature of our geography dictates that we're going to continue being a bedroom to Houston," said Moore.

The rapid growth in Conroe continued to create customer problems, as reflected by an August 19, 1977, letter to *The Conroe Courier* by a writer who complained "that the current nuisance...is the fading in and out of the caller's voice, which makes you think someone is playing with a volume control somewhere."

In response to the community's criticism of their work, more than two dozen Conroe telephone employees wrote to the editor: "...we are a highly trained and dedicated group of workers who work hard every day to provide this community with telephone service. Our day doesn't always end at five o'clock because we are all subject to 24-hour call. If a line is torn down, a cable cut or a toll carrier goes down, we are called to get it fixed regardless of the hour..."

Conroe subscribers also complained about having to pay toll charges to Willis, eight miles away. Conroe and Lufkin telephone officials tried to buy the Willis exchange in the 1970's, but the owner refused to sell. General Telephone today owns the Willis system.

The year 1977 also saw Morris Bateman, whom *The Conroe Courier* called "the man who lives with the daily complaints about his phone company," named the Courier's "Citizen of the Year" at the 41th annual banquet of the Chamber of Commerce. "This is one for the books," joked Bateman, "with as many enemies as I have working with the telephone company. I hope all of your phones are working when you get home tonight."

Introducing Bateman, Courier Publisher Floyd Garrett praised him for his work with the Red Cross, the United Fund, the Montgomery County Day Camp, the Conroe Noon Lions Club, the Boy Scouts, the First United Methodist Church, and the Chamber of Commerce.

Ironically, a few weeks later, Conroe's telephone service was disrupted when a piece of underground equipment was damaged, knocking out the Courier's Associated Press news wire and delaying production of the newspaper.

In November of 1977, telephone workers at Conroe, Lufkin and Alto ratified a new three-year wage agreement, averting a strike. The workers reversed themselves from an earlier decision when they voted 75 to 25 percent to reject the first offer of the company's management. The ratifying vote was 60 to 40 percent. The new contract called for an average 29.9 percent pay and fringe benefit increase for union workers over the three-year period.

At Ross' urging in 1978, the Board of Directors of Lufkin-Conroe Communications Co. created a new wholly-owned subsidiary, Lufkin-Conroe Telecommunications Corporation, to sell telephone equipment to its customers, an opportunity he predicted would be opened up by the deregulation of customer premise equipment. The move represented a change from the days when telephone companies owned everybody's phones.

In 1978, still looking for a technology that would replace the traffic-sensitive electromechanical switching system of handling telephone calls, Ross traveled to Florida to attend a meeting sponsored by Northern Telecom to announce "The Digital World."

"We had known that digital switching was coming, but this was the first time that a digital central office seemed practical and technically possible," said Ross. "What we saw in Florida was the answer to all of our problems in Conroe and Lufkin."

In 1978, Ross negotiated an agreement with Northern to install digital switching in Conroe by 1980 and in Lufkin by 1982. "Northern's regional sales people came to Lufkin, we spent a week working on the plan, and bought a package for $14 million," he said.

Although some digital systems had been installed in Canada, the Conroe installation of a Northern Telecom DMS-100/200 digital switching system became the first in the United States. By 1980, the company completely converted its Conroe operations to a new 18,000-line computerized digital system, one of the most technologically advanced switching systems in the world.

A new 5,000-square foot wing was added to the Phillips Street address to make way for the digital system, and all of the old switching gear was sold to a telephone salvage company. "They came in here with three huge trucks and hauled it all off," recalls Bateman.

While the digital switching system came to Lufkin in 1982, it actually came to a nearby community, Diboll, several years earlier. Diboll was chosen for the system by the company in 1978, said Ross, because it was the proper size to accommodate the digital equipment that was available. Ross said while the technology to build the new digital switching facilities was available, a complex computerized system to serve a city the size of Lufkin was not available. By 1982, the technology had developed sufficiently to allow the company to convert the Lufkin system to digital switching.

The high-tech digital system enabled the Conroe and Lufkin exchanges to switch calls almost instantly, ending traffic congestion, busy signals, and most service problems. "It was the end of most of our headaches," said Ross.

The digital program was completed at Riverbrook on January 15, 1979, came to Lake Conroe on April 23 the same year, and expanded to Walden on May 19. Conroe's 539 exchange received the digital service on June 19, 1980, and the 756 exchange received it the following month.

The switching system gave the exchange an opportunity to increase its efficiencies by "100 times," according to Bateman in a newspaper quote.

The late 1970s also saw the completion of a new service center on Old Magnolia Road in Conroe. "It was something we needed badly, but the people living there didn't want to sell any of their property. Finally, we persuaded one of the owners, who lived in Chicago, to sell, and the others went along," remembers Bateman.

Ross said the digital systems might not have been possible without a Lufkin banker, H.J. (Bubba) Shands, the nephew of Ned Shands, a son-in-law of telephone pioneer Alexander Madison Denman. "We went to a Conroe banker, and wanted to borrow some money, but he declined to make the loan, saying the bank's core business was widow women and demand accounts," he recalls. "But we went to First Bank and Trust, and Bubba showed an interest in working with us."

The company sought $14 million in long-term debt during an era of double-digit inflation and high interest rates, but Shands suggested a short-term loan at his bank until long-term rates came down. "It made a difference, we saved a lot of interest costs, and Bubba's help was essential," said Ross.

In 1982, as Ross had promised the community of Lufkin, a Northern Telecom DMS 100/200 digital switching system was placed in service, and the Alto community received a DMS-110 digital switching system in 1986, making all of the company's exchanges 100 percent digital.

Uniformly throughout the company's network, consumer complaints plummeted. Conroe and Lufkin city officials praised the company for its work. The company entered a new era of technology which began to open up new telecommunications opportunities.

At about the same time, however, Ross and other utility managers in Texas had a new problem on their hands: the creation of the Texas Public Utility Commission (PUC). Ironically, the regulatory agency was brought about by Lufkin's then state senator, Charlie Wilson, who carried on a running feud with American Telephone and Telegraph Company and other utility companies over high rates.

The Historical and Creative Arts Center of Lufkin, now known as the Museum of East Texas, was featured on the 1977 Lufkin telephone directory cover.

Wilson had begun campaigning for utility regulation in Texas as early as 1969. Writing in *The Diboll Free Press* in May of 1969, Wilson claimed that Texas utility companies "are the most profitable in the country, and that Texas consumers pay the highest rates in relation to the cost of living of any consumer in the United States."

At the height of his campaign, Ross and Marion Clay showed up at a Wilson reception at the Alabama-Coushatta Indian Reservation and presented the state senator with a special telephone. Throughout his campaign for utility regulation, however, Wilson never complained about the Lufkin-Conroe telephone service and even held it up as a model for others in the business.

The Public Utility Commission, created by a law passed in 1975 and effective in 1976, forced utility companies to file rate cases with the agency in Austin. In December of 1983, Ross filed with the PUC a rate case for Lufkin. The company has never found it necessary to return to Austin with another rate increase.

Board members of Lufkin Telephone Exchange and Conroe Telephone Company in the 1970's were B.H Denman, H.T. Bowers Sr., N.D. Shands Jr., Peyton L. Denman, Marion Clay, E.J. Bowers, James Newman, Kester W. Denman, Jr., Edwin M. Simons, Frank R. Denman, Darrell Hamric, Shirley Simons, Jr., Morris Bateman, Douglass Shands, Ned D. Shotwell and G.I. Ross.

Lufkin lineman Gerald Freeman checks a line in the 1980's.

18 "One family and one company"

By the early 1980s, most of the customer complaints about poor telephone service in Conroe and Lufkin had ended. The electromechanical method of switching subscriber calls and handling the flow of telephone traffic had been replaced by a digital system with almost none of the moving parts that made the old system vulnerable to failures.

As the first digital telephone system in the United States, the Lufkin-Conroe system gained a new respect, not only among its customers, but among other telephone companies. The growth of the Conroe system, for example, resulted in 22,814 working telephones by March 31, 1981, an increase of nearly 12,000 in 10 years. By the end of 1981, Conroe had 24,666 working telephones.

The turnaround enabled the Mantooth and Denman families, Larry Ross, and his staff to focus on new opportunities.

One opportunity lay with the company's employees. "With the advent of the digital technology, the expertise and knowledge of our employees had to be raised, not only in maintenance and engineering, but in many other areas of telephone management," recalls Ross. "We began taking steps to develop a very skilled maintenance department, not only with training, but with selective hiring."

The company pulled most of its new engineering and marketing personnel from Texas A&M University. "A&M has given us some of the brightest and best people in the company, and over the years, we've depended on the university a great deal," said Ross.

The original home of the Lufkin telephone exchange before the building was remodeled in the late 1980s. The telephone company was located on the building's second floor.

In 1980 at Conroe, Ross announced the promotion of three key managers. Morris Bateman, vice-president and general manager at Conroe, was promoted to executive vice-president, and customer service manager Numsen M. Hail was named general manager of the exchange.

Doris Johns, who

Larry Ross, right, presents a check for $5,000 to Angelina College President Jack Hudgins to help fund construction of a new health careers building at the college.

worked in Conroe until her retirement in 1986, said the efforts by Ross and other company executives in the 1980s "did a lot to bring the company together as one family and one company." Before Ross' arrival, she said, Lufkin and Conroe operated "like they were separate companies." As time passed, the company became better organized and more efficient as a single company, she said.

In September of 1982, death came to Marion A. Clay, who had retired in 1974 but remained on the board of directors of Lufkin-Conroe Communications as an advisory director. He worked with the Boy Scouts, was secretary-treasurer and also served as president of the Lufkin Lions Club, was a member of the board of directors of the Salvation Army, and was a member and deacon of the First Baptist Church where he had served on numerous committees.

He had also served as director of the Texas Telephone Association, as well as president and chairman of nearly all of the associations committees. He served as a director of the United States Independent Telephone Association. In 1979, Mr. Clay attended his 50th class reunion at Texas A&M University.

Lufkin's 1980 telephone directory depicted a scene around Rayburn Lake in Deep East Texas.

A series of federal and state regulatory actions in the 1980's and 1990's gave the company new opportunities.

The deregulation of customer premise equipment in the late seventies, while

Employees of the Lufkin telephone exchange posed for this photograph in 1980. Left to right are Johnie Weatherford, Marvin Schmidt, Robert Brock, Jack Mott, Robert Beavers, Gerald Glover, Ronnie Davis, Emory Quarles, Monroe Ivy, J.B. Moore, Charlie Lawrence, Chuck Lundstedt, Bill Bowling, Walter Lawrence, Oval Terry, Charles Redden, Toby Appleman, Al Schuchardt, Dale Green, Joe Noel, Dean Hamilton, Bob Mason, Linda McLeroy, Ken Stone, Tom Wetterman, Margie Tindall, G.I. Ross, Bob Katzenberger, Debbie Barnes, Jim Swor.

The 1981 officers of Lufkin-Conroe Communications, Conroe Telephone Company, Lufkin Telephone Exchange, and Alto Telephone Company were, standing left to right, E.J. Bowers and Peyton L. Denman. Seated, left to right, are G.I. (Larry) Ross, Shirley Simons, Jr., and Kester W. Denman, Jr.

representing a culture shift from the days when telephone companies owned everybody's phones, enabled the company to emerge as a retailer of telephone equipment throughout Southeast Texas. "We had to adapt to the change, but we learned quickly and soon found we could prosper in the new environment," said Ross.

In April of 1980, the company announced in a newspaper ad: "We are now offering telephone equipment for sale. Even Key Telephones and PBX Systems...you can now purchase telephones you had to pay monthly rent for in the past. Federal legislation has been aiming at opening up the communications industry to competition for years and most recently has required telephone companies to sell telephones under

The Conroe telephone exchange completed this building on Frazier Street in the 1980s as part of its upgrading of the exchange's facilities.

The 1981 directors of Lufkin-Conroe Communications, Conroe Telephone Company, Lufkin Telephone Exchange, and Alto Telephone Company were standing, left to right, Morris Bateman, Sr., Douglass Shands, Frank R. Denman, James E. Newman, Peyton L. Denman, and Ned D. Shotwell. Seated, left to right, are Edwin M. Simons, G.I. (Larry) Ross, Shirley Simons, Jr., Kester W. Denman, Jr., E.J. Bowers, and Linwood D. Newman.

Lufkin's North First Street telephone offices looked like this in the 1960's before extensive renovations took place.

FCC Order 19528. There is no mistaking the intent of current legislation..."

In July of 1985, Lufkin Telephone Exchange announced that it would take over sales and service of business phone systems from its sister company, Lufkin Conroe Telecommunications, headquartered in Houston. "Last week, if a customer wanted to rent a phone, he'd have to go to the LCT office located in our building, All paperwork on sales and service had to be kept separate. It's been a nightmare to keep all these records. Now, the customer can buy, rent or lease directly from LTX and pay less," said vice-president Charles Redden in a company statement.

One of the company's earliest efforts was to focus on large business systems and, as a result, the company sold a multi-million dollar contract to the General Services Administration in El Paso and another to the Houston Transit System. "For a while, we reached out to a lot of cities beyond our traditional service area, but it was too much effort for too little business, so we moved back to East Texas," said Dale Green, today's president of LCTX, Inc.

Between 1980 and 1985, the members of the board of directors for Lufkin Telephone Exchange and Conroe Telephone were Darrell Hamric, Linwood D. Newman, Peyton L. Denman, Edwin M. Simons, Kester W. Denman, Jr., Marion Clay, E.J. Bowers, James E. Newman, G.I. Ross, Frank R. Denman, Shirley Simons, Jr., Douglass Shands, Ned D. Shotwell, Morris Bateman, and Chester H. Moore.

In 1986, the boards took steps leading to the merger of the three existing telephone companies—Lufkin, Conroe and Alto—into a single corporation, Lufkin-Conroe Telephone Exchange, Inc. (LCTX), to allow the consolidation of

Directors of Lufkin-Conroe Communications and Lufkin-Conroe Telephone Exchange in 1985 were, standing left to right, James E. Newman, Linwood D. Newman, Chester H. Moore, Peyton L. Denman, and Ned D. Shotwell. Seated, left to right, are Edwin M. Simons, G.I. (Larry) Ross, Shirley Simons, Jr., Kester W. Denman, Jr., and E.J. Bowers.

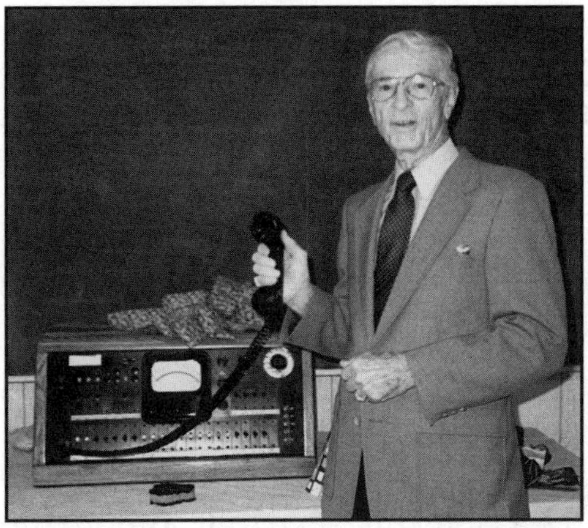

At his retirement party in 1985, Conroe exchange officer Morris Bateman was honored for 47 years with the company. He was given this 1954 dial test board.

the ownership. Prior to the reorganization, the three companies had been operating independently. The merger, coupled with a federal tax reform act in 1986, produced a rate reduction for customers within the system.

The new Lufkin-Conroe Telephone Exchange, Inc. was governed from 1986 to 1990 by a board made up of E.J. Bowers, William P. Bowers, Calvin W. Jayroe, Frank R. Denman, Kester W. Denman, Jr., Peyton L. Denman, Chester H. Moore, James E. Newman, Linwood D. Newman, G.I. Ross, Douglass Shands, Edwin M. Simons, Ned D. Shotwell, Shirley Simons, Jr., and Thomas O. Shelton, III.

The officers of the company were Ross, chairman, president and chief executive officer; Frank R. Denman, secretary; Dale Green, who served as vice-president, treasurer, assistant secretary, and executive vice-president; Numsen Hail, vice-president; Ken Whitton, vice-president of revenues; Deborah Huffman, controller, treasurer and assistant secretary; and Kirk Ross, assistant vice-president.

The 1985 Lufkin telephone directory focused on Angelina, the Indian maiden for whom Angelina County was named.

The year 1986 brought about the expansion of the company's Lufkin complex with the addition of a new 8,400-square-foot building located adjacent to the corporate offices on North First Street. The new $360,000 structure, built at the corner of North First and Groesbeck, was designed to house commercial operations of the company, including the cashier and business offices and phone sales, according to Redden. "The purpose of the addition is to provide better service to customers in the surrounding area," he said.

In 1985, the origin of much of the Lufkin telephone company's early history saw a change in its appearance. The old Denman Building, where Lufkin's first switchboard system was located in 1900, was remodeled by its owner, R.H Duncan Construction Company. The remodeling was completed to house several professional offices. Later, when the Lufkin exchange acquired the property, a parking lot was built to serve the building. In the process, a next-door building owned by the D.H. Townsend family was demolished. The building once housed a furniture store, bicycle shop, and bookstore. The Denman building was originally built around the turn of the century and remodeled by the Denman family in 1939.

The decade of the eighties also saw the creation of the Texas IntraLATA Toll Pool, which was organized by the Texas Public Utility Commission to divide intrastate toll revenues and inter-exchange carrier access charges; the expansion of the Lufkin telephone directory to include listings from several surrounding communities, including Nacogdoches, Chireno, Huntington, and Zavalla; and the development of the 9-1-1 emergency system in the Lufkin/Conroe areas, although actual service did not begin until the early 1990s.

In the early 1990s, the court-ordered breakup of American Telephone and Telegraph Company (AT&T), which was effective in 1984, brought additional changes. "We didn't know what the fallout would be, but we soon learned to take advantage of the new landscape," said Ross.

One area in which the company prospered was long distance tolls. Before the AT&T breakup, Lufkin and Conroe divided revenues from long distance toll calls with the Bell system on roughly an 80-20 basis. "Every time someone in Lufkin made a long distance call to Houston, we got 80 percent of the revenue and Bell got 20 percent, but if a call originated in Houston, Bell kept 100 percent. The divestiture changed the rate structure and improved our retention of toll charges."

The AT&T breakup also led to the creation of Local Access Transport Areas (LATAs), which also changed the way the company charged its customers for long-distance calls. "It meant additional administrative costs, but it meant some additional revenue, too," said Dale Green.

Into the telephone industry came new long-distance players like MCI and Sprint, which began to grow, compete against existing long-distance carriers, and to carve out a place in the country. But the new long-distance providers had little impact on the Lufkin and Conroe systems. "Again, we incurred more administrative costs to keep up with long-distance charges, but as MCI and Sprint forced down the costs, there were a lot more calls, and we benefited," said Green.

In 1990, the company upgraded its Northern Telecom digital switches to Supernodes in the Lufkin and Conroe central offices, the first step in providing Centrex (an alternative to PBX's), custom local area signaling services, and other advanced features.

Officers of Lufkin-Conroe Communications and Lufkin-Conroe Telephone Exchange in 1986 were, standing left to right, Dale Green, Deborah Huffman, Peyton L. Denman, and Kenneth R. Whitton. Seated, left to right, are G.I. (Larry) Ross, Shirley Simons, Jr., Kester W. Denman, Jr., and E.J. Bowers.

In 1989, G.I. (Larry) Ross was honored with the Texas Telephone Association's Nevile Haynes Award for a lifetime of leadership and service to the telephone industry.

"The old PBX systems we installed before the digital technology of the 1980s were not always customer friendly, but the digital Centrex service of the mid 1990s represented a big improvement, so we aggressively moved into this market," said Green.

The improved digital technology also meant the introduction of a host of new features for the company's customers, including selective call forwarding, distinctive ringing, and caller identification. The company was among the first local exchange carriers in Texas to provide the Caller ID option.

Green feels one of the company's greatest technology leaps was the installation of fiber optic lines. "Following the opportunities opened up by the divestiture of AT&T, our move to fiber optics permitted us to lower our maintenance costs, handle a greater number of calls, and move into some joint ventures with other companies."

In one such venture, LCT built a fiber optic line from Lufkin to Madisonville, connecting with a Dallas-to-Houston line. Another fiber optic line was built from Abilene to Goldwaithe, a third was constructed in the Longview area, and a fourth was built from Lufkin to the Houston area in cooperation with East Texas Telephone Cooperative. "Fiber optics did a lot to open up new territory for us," said Green.

The fiber optics technology not only allows more calls and an improved environment for transporting data and video transmissions, but limits "ditch opening," the most costly element in line maintenance. "In theory, you should never have to open a fiber optics ditch unless someone cuts into it," said Green.

The 1980's also brought an era to a close. Kester W. Denman, Jr. who was responsible for hiring Ross, retired on April 1, 1987, as vice-president and general counsel for the company. He had been an employee since 1974 when he retired from an earlier legal position at Southwestern Bell in Houston.

Shortly after World War II, the Denman and Mantooth families had adopted an informal policy that no family members would work for the company. Kester Denman excepted from this policy due to his extensive legal experience in the telephone industry.

"Ned D. Shands, Jr., a member of the board, told me when I joined the company that the family had decided that hiring family members wasn't good policy because every time you promoted a relative, it created problems, not only among family members, but other employees as well," said Ross. "The board wanted professional telephone people to run the company."

The decision earned praise more than once from the Texas Public Utilities Commission. "They checked us on this issue carefully, and they decided that we were, indeed, a professional utility," said Ross.

In 1995, another regulatory decision by the Texas Legislature provided for increased competition among local telephone exchanges, including LCTX. When the Texas Public Utility Regulatory Act of 1995 became effective in 1996, it set in motion a series of steps that could eventually impact LCTX services.

Telephones began to take on a new appearance in the 1980's, as evidenced by this Mickey Mouse phone.

The law allows for LCTX to sell telephone service to another supplier, which could, in turn, re-market the service to customers, even in the same town. It also permits other telephone providers to use the LCTX network by paying for the infrastructure and maintaining a portion of the network. It allows new telephone companies to do everything LCTX does in the same community.

While the total impact of the new law has yet to be determined, Green and Ross feel it may offer opportunities for the company "if we do our job well and keep our customers satisfied." Because of LCTX's city franchises, Ross feels Lufkin and Conroe would be reluctant to scrap the LCTX investment in the communities.

Green, like others in the company, feels the new law presents a number of unanswered questions, including the maintenance of the 9-1-1 emergency call program and the retention of the customer's traditional phone number. "Just how fast competition comes to the local exchanges, like Lufkin and Conroe, is anyone's guess," said Green.

By the close of 1997, LCTX was ranked as the fourth largest local telephone exchange in Texas and the 20th largest non-Bell company in the U.S. Its service area extended over 1,400 square miles with approximately 4,560 miles of telephone cables providing service to over 97,000 access lines.

The directors of the Lufkin-Conroe Telephone Exchange, Inc. in the decade of the 1990s were William P. Bowers, Thomas O. Shelton, III, Calvin W. Jayroe, Frank R. Denman, Chester Moore, Linwood D. Newman, G.I. Ross, Douglass Shands, Edwin M. Simons, Ned D. Shotwell, Shirley Simons, Jr., Peyton L. Denman, Stephen P. Moore, Carlos R. Hamilton, Deane C. Watson, Jr., Suzanne Simons Laidlaw, Douglas S. Shotwell, and Dale L. Trimble.

Officers of the company during the 1990s were Ross, president, chairman and chief executive officer; Frank R. Denman, vice-president, and Stephen P. Moore, secretary; Dale Green, executive vice-president and president; Numsen Hail, vice-president and vice-president of public affairs; Ken Whitton, vice-president, senior vice-president and executive vice-president; Kirk Ross, assistant vice-president, vice-president of operations, and senior vice-president; and Deborah Huffman, vice-president, controller, treasurer, and assistant secretary; Mike Robinson, vice-president of marketing and planning; and Russell Buras, assistant vice-president.

At the close of 1997, the directors and officers of Lufkin-Conroe Communications Co., the parent company, were Linwood D. Newman, chairman; Suzanne Simons Laidlaw, Douglass Shands, William P. Bowers, Thomas O. Shelton, III, Calvin Jayroe, Stephen P. Moore, Carlos Hamilton, Jr., Deane C. Watson, Jr., Douglas S. Shotwell, Dale L. Trimble, and G.I. Ross. Officers were Ross, president and chief executive officer; Dale Green, executive vice-president; Deborah Huffman, vice-president, controller, treasurer and assistant secretary; Ken Whitton, executive vice-president; Kirk Ross, senior vice-president; and Stephen P. Moore, secretary. G.I. (Larry) Ross was also chairman and president of Lufkin-Conroe Telecommunications Corporation, a wholly-owned subsidiary.

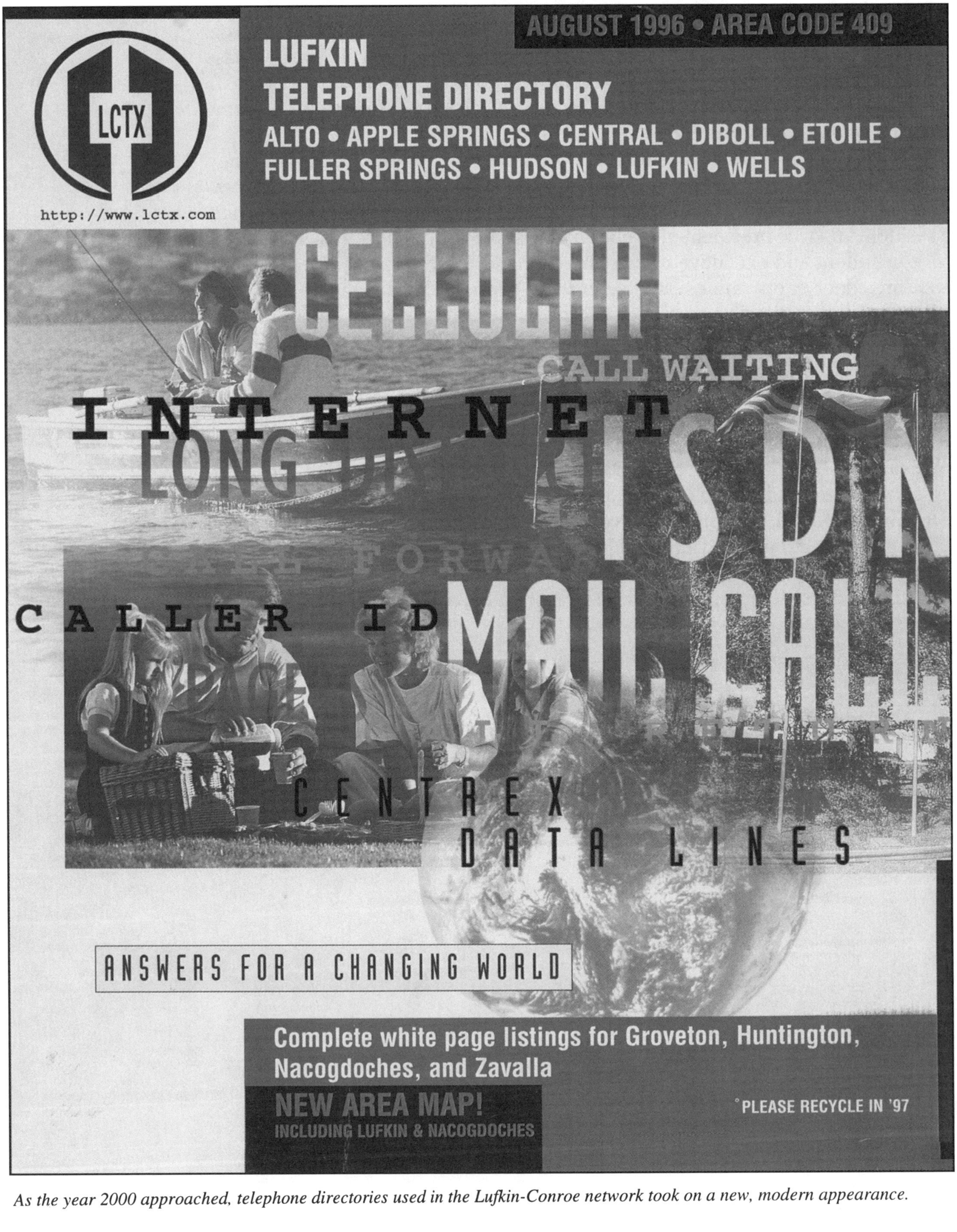

As the year 2000 approached, telephone directories used in the Lufkin-Conroe network took on a new, modern appearance.

19 A profile of leadership

Only seven individuals have served as president of the Lufkin and Conroe telephone exchanges and its subsidiaries since the 1890s.

Dr. Alexander Madison Denman, who founded the company in 1898 with Judge Edwin James Mantooth, was president of the Lufkin Telephone Company from its inception until his death in a 1908 automobile accident. In 1909, Judge Mantooth became president and remained in the role until his death in 1944.

In 1944, Kester W. Denman, Judge Mantooth's law partner and Dr. Denman's son, took over the presidency until 1948 when his brother, Dr. L.H. Denman, was named to the position.

Dale Green was elected president of Lufkin-Conroe Telephone Exchange, Inc., in 1996.

When the company's name was changed to Lufkin Telephone Exchange in 1957, Dr. Denman continued as president until his death in 1968.

In 1968, the company's board of directors named Marion Clay the first president to hold the office outside the Denman or Mantooth families. Morris Bateman served as manager of the Conroe company in August of 1948, but was not an officer. He was promoted to vice-president in 1969 and became executive vice-president in 1981. He remained in this post until he retired in December, 1985

Clay remained president until May, 1973, when G.I. (Larry) Ross was promoted from executive vice-president to the top office. Ross continued as president until March of 1981 when he became president and chief executive officer. Serving as operating officers with Ross in 1981-1985 were Charles Redden, vice-president; and Dale Green, vice-president, secretary and treasurer.

When the Lufkin-Conroe Communications Co. was created in 1969 as the parent company for the Denman and Mantooth telephone interests, Clay became its first president, but relinquished the presidency six days later to Ned D. Shands, Jr. Clay returned to the presidency in September of 1970 and continued his responsibilities until Ross was elected president in May of 1973.

Ross was promoted to president and chief executive officer of LCC in March of 1981 and remained in the position at the end of 1997. Other LCC officers included Dale Green, executive vice-president, vice-president, treasurer, assistant secretary; Deborah D. Huffman, vice-president, controller, treasurer, and assistant secretary; Kenneth R. Whitton, executive vice-president; and Kirk Ross, senior vice-president.

An early Lufkin-Conroe board of directors meeting included, left to right, Ned D. Shands, Jr., Shirley Simons,Jr., Kester W. Denman, Sr., Harry T. Bowers, Linwood H. Denman, Byford H. Denman, Frank R. Denman, and Kester W. Denman, Jr.

In 1986, the Lufkin and Conroe exchanges were merged into a single company, the Lufkin-Conroe Telephone Exchange, Inc., and G. I. (Larry) Ross was elected president and chief executive officer of the new company. In March of 1987, he was named president, chairman and chief executive officer. Dale Green was named president in December, 1996. They continued their duties at the end of 1997.

Green also served LCTX as executive vice-president, vice-president, treasurer and assistant secretary; Charles Redden was a vice-president. Numsen Hail was a vice-president and later vice-president of public affairs. Kenneth R. Whitton was a vice-president, vice-president of revenues, senior vice-president, and executive vice-president. Deborah D. Huffman was a vice-president, controller, treasurer, and assistant secretary. Kirk Ross was a vice-president, vice-president of operations, and senior vice-president. Michael B. Robinson was vice-president of marketing and technology.

A similar succession of presidents occurred in the Conroe exchange, which, like the Lufkin exchange, remained a separate company until 1986.

Judge Mantooth became president of the Conroe system in 1929 when it was

Directors of Lufkin-Conroe Communications and Lufkin-Conroe Telephone Exchange in 1992 were, front row, left to right, Linwood D. Newman, Shirley Simons, Jr., G.I. (Larry) Ross, Peyton L. Denman; back row, left to right, Thomas O. Shelton, III, Carlos R. Hamilton, Jr., William P. Bowers, N. Douglass Shands, III, Stephen P. Moore, Calvin W. Jayroe, Edwin M. Simons, and Ned D. Shotwell.

purchased from George B. Madeley and incorporated as the Conroe Telephone Company. He continued in the role until his death. He was then succeeded by Kester W. Denman, Sr. Linwood H. Denman took over the company's leadership in 1948, continuing until his death.

In April of 1968, Ned D. Shands Jr. became chairman of the Conroe operations and Marion Clay became president. G.I. Ross succeeded Clay in 1973 and remained president until his title was changed to president and chief executive officer in March of 1981.

Before the merger, other officers of the Conroe exchange were Morris Bateman, executive vice-president; Charles Redden, vice-president; Dale Green, vice-president and treasurer, and Numsen Hail, vice-president and general manager.

When the Mantooth and Denman families purchased the Alto Telephone Company in March of 1963, L. D. Denman became its president. He was succeeded by Marion Clay in April, 1968, and Ross succeeded Clay as president in May, 1973. Ross continued in the office until the exchange was merged into the Conroe and Lufkin systems in 1986. Other officers before the merger were Charles Redden, vice-president, and Dale Green, vice-president and treasurer.

Since the inception of telephone service in Lufkin, a number of Denman and Mantooth family members have served as directors of the Lufkin Telephone Exchange, the Conroe Telephone Company, the Lufkin-Conroe Telephone Exchange, the Alto Telephone Company, and Lufkin-Conroe Communications Co.

Between 1898 and 1985, the directors of the Lufkin Telephone Exchange included Dr. Alexander Madison Denman, Judge Edwin James Mantooth, Mrs. A.M. Denman, Shirley Simons, Sr., Mrs.Ned D. Shands, Mrs. Dell Lanter, Ned D. Shands, Sr., Thomas O. Shelton, Jr., Peyton R. Denman, N. Douglass Shands III, Linwood H. Denman, Byford H. Denman, Kester W. Denman, Sr., Frank R. Denman, Harry T. Bowers, Sr., Denman C. Hucherson,

Officers of Lufkin-Conroe Communications and Lufkin-Conroe Telephone Exchange in 1992 were, left to right, Michael Kirk Ross, Kenneth R. Whitton, Peyton L. Denman, Shirley Simons, Jr., G.I. (Larry) Ross, Deborah D. Huffman, Dale Green, and Numsen M. Hail.

Ned D. Shands, Jr., Kester W. Denman, Jr., Shirley Simons, Jr., Darrell Hamric, Peyton L. Denman, Marion Clay, Ned D. Shands, Jr., Edwin J. Bowers, James E. Newman, G.I. Ross, Edwin M. Simons, Linwood D. Newman, and Jean Shotwell.

Between 1929 and 1986, the directors of the Conroe Telephone Company included Edwin J. Mantooth, Ned D. Shands, Sr., Linwood H. Denman, Denman C. Hucherson, Mrs. A.M. Denman, Kester W. Denman, Sr., Shirley Simons, Sr., Jean Shotwell, Harry T. Bowers, Peyton R. Denman, Ned D. Shands, Jr., Mrs. Ned D. Shands, Mrs. Dell Lanter, Thomas O. Shelton, Jr., Frank R. Denman, Byford H. Denman, Shirley Simons, Jr., Kester W. Denman, Jr., Darrell Hamric, Peyton L. Denman, Marion Clay, Edwin J. Bowers, Morris Bateman, G.I. Ross, N. Douglass Shands, III, Ned D. Shotwell, and Chester Moore.

Morris Bateman of Conroe was inducted into the Independent Telephone Pioneer Association's Hall of Fame in the 1990s. He served as president of the Texas Pioneer Chapter as director, vice-president, and president. He was also chairman of the Hall of Fame. He holds an ITPA life membership.

Between 1963 and 1986, the directors of the Alto Telephone Exchange included Linwood H. Denman, Frank R. Denman, Kester W. Denman, Sr., Denman C. Hucherson, Shirley Simons, Jr., Darrell Hamric, Edwin J. Bowers, Marion Clay, Peyton L. Denman, Harry T. Bowers, Sr., Byford H. Denman, James E. Newman, G.I. Ross, Ned D. Shands, Jr., Kester W. Denman, Jr., Edwin M. Simons, and Linwood D. Newman.

Between 1986 and 1997, the directors of the Lufkin-Conroe Telephone Exchange, Inc., were Edwin J. Bowers, Frank R. Denman, Kester W. Denman, Jr., Peyton L. Denman, Chester Moore, James E. Newman, Linwood D. Newman, G.I. Ross, N. Douglass Shands, III, Edwin M. Simons, Shirley Simons, Jr., William P. Bowers, Thomas O. Shelton, III, Ned D. Shotwell, Calvin W. Jayroe, Carlos R. Hamilton, Jr., Stephen P. Moore, Douglas S. Shotwell, Deane C. Watson, Jr., Suzanne Simons Laidlaw, and Dale L. Trimble.

Between 1969 and 1997, the directors of the Lufkin-Conroe Communications Co. were Ned D. Shands, Jr., Marion Clay, Byford H. Denman, Frank R. Denman, Kester W. Denman, Jr., Peyton L. Denman, Edwin J. Bowers, Shirley Simons, Jr., Darrell Hamric, G.I. Ross, Harry T. Bowers, Sr., James E. Newman, Edwin M. Simons, N. Douglass Shands, III, Ned D. Shotwell, Chester Moore, Linwood D. Newman, Calvin Jayroe, Thomas O. Shelton, III, Carlos R. Hamilton, Jr., Stephen P. Moore, Deane C. Watson, Jr., William P. Bowers, Douglas S. Shotwell, Dale L. Trimble, and Suzanne Simons Laidlaw.

G.I. Ross has served as chairman of Lufkin-Conroe Telecommunications, a company subsidiary, since it was created in 1978.

CHRONOLOGY OF THE DENMAN AND MANTOOTH FAMILIES

(Represented by membership on the Lufkin-Conroe Communications Co. board of directors in 1997)

William P. Bowers, Irving, Texas
Father: Edwin J. Bowers, Sr.
Grandmother: Robbie Mantooth Bowers
Great-grandfather: Edwin James Mantooth

Dr. Carlos R. Hamilton, Jr., Houston, Texas
Mother: Bertha Denman Hamilton
Grandfather: Dr. Peyton R. Denman
Great-grandfather: Dr. Alexander Madison Denman

Calvin W. Jayroe, The Woodlands, Texas
Mother: Rhoda Shotwell Jayroe
Grandmother: Winnie Mantooth Shotwell
Great-grandfather: Edwin James Mantooth

Suzanne Simons Laidlaw, Dallas, Texas
Father: Edwin M. Simons
Grandmother: Mollie Mantooth Simons
Great-grandfather: Edwin James Mantooth

Stephen P. Moore, Lufkin, Texas
Mother: Caledonia Denman Moore
Grandparents: Kester W. Denman, Sr. and Cleo Mantooth Denman
Great-grandfathers: Dr. Alexander Madison Denman and Edwin James Mantooth

Linwood D. Newman, Houston, Texas
Mother: Mary Claire Denman Newman
Grandfather: Dr. Linwood H. Denman
Great-grandfather: Dr. Alexander Madison Denman

N. Douglass Shands, III, Austin, Texas
Father: Ned D. Shands, Jr.
Grandmother: Lillian Denman Shands
Great-grandfather: Dr. Alexander Madison Denman

Dr. Thomas O. Shelton, III, St. Louis Park, Minnesota
Mother: Jane Denman Shelton Hamric
Grandfather: Dr. Linwood H. Denman
Great-grandfather: Dr. Alexander Madison Denman

Douglas S. Shotwell, Houston, Texas
Father: Ned D. Shotwell
Grandfather: Jean Shotwell
Great-grandmother: Winnie Mantooth Shotwell
Great-great-grandfather: Edwin James Mantooth

Dale L. Trimble, The Woodlands, Texas
Father-in-law: Dr. Peyton L. Denman
Grandparents-in-law: Kester W. Denman, Sr. and Cleo Mantooth Denman
Great-grandfathers-in-law: Dr. Alexander Madison Denman and Edwin James Mantooth

Deane C. Watson, Jr., Plano, Texas
Mother: Mollie Simons Watson
Grandmother: Mollie Mantooth Simons
Great-Grandfather: Edwin James Mantooth

The Lufkin telephone exchange modernized its North First Street facilities in Lufkin during the 1980's.

In the 1990's, the Conroe telephone exchange moved into this modern building just off Interstate 45 in Southwest Conroe. Many of the exchange's operations were consolidated in the building.

20 Heading into a new century

In 1898, when Dr. Alexander Madison Denman and Judge Edwin James Mantooth made a telephone call to each other, they picked up an instrument in their office, turned a crank, and rang another phone a few blocks away.

The voices traveling over the overhead wires were sometimes scratchy and difficult to understand. Long distance service had not yet arrived in Lufkin and the telephone was still considered a novelty.

In 1998, when Cori Jo Bowers, 21, a great-great-great granddaughter of E.J. Mantooth, uses the telephone, it may be to call Lufkin, Little Rock or London and the voices may be as clear as those of a neighbor. It may not even be a call in the traditional sense. It could be the catalyst for dialing into a world-wide network of information and services, to order a product from a mail-order house in Maine, arrange a conference of friends or business associates, or leave an E-mail, voice mail, or video message with a friend in Canada.

In 100 years, the shape of the telephone enterprise created by Dr. Denman and Judge Mantooth has changed so much that the two Lufkin pioneers would have difficulty believing what they wrought when they purchased a pair of simple crank telephones in 1898.

But, if telephone technology moves as fast as it did in the 1980s and 1990s, the leaps of the past century could be eclipsed in a decade or less.

In the May-June, 1997, issue of *The Connection*, a publication of Lufkin-Conroe Communications Co., G.I. (Larry) Ross, president and chief executive officer, told LCC employees:

"With the communications industry in the midst of change, opportunities and challenges are abundant.

"Our company has faced competition in many segments of its business for several years. Now, we are on the threshold of confronting competition in our local loop.

"Our future depends on how you and I respond to these opportunities each day we come to work. Looking toward the challenges ahead, we must accomplish three main objectives: solid customer service, continue to strengthen our network, and sharpen our skills, especially in the data and video side of the business.

"First, our greatest asset is our customer base. We must recognize that our level of customer service is a large part in acquiring and retaining our customers' business. Customer service is critical to remain competitive.

"Second, we have a huge investment in our network. We must be diligent in our efforts to protect this investment by ensuring its continued capability to deliver the services our customers need.

"Third, each of us has an investment in this company. From accounting to regulation, from customer service to technology, our world is changing. The new knowledge and the new skills our people must develop will be critical to our future service.

"We all understand our business is changing, but more importantly, I have great confidence in our employees' ability and their 'we can do it' attitude. Each of you should want to be a part of this change.

"Let us all work together to continue to deliver the best communication services to our customers."

As the 1990s drew to a close, Ross and the more than 500 men and women working for the company continued to make changes positioning LCC for the next decade.

Looking at the company's future, Michael B. Robinson, LCTX vice-president of marketing and technology, sees far-reaching communications developments that will affect business and residential telephone subscribers.

"More and more, our industry is moving to network-based services that will allow business intelligence to be derived from other sources, rather than just a single box in a business office," he said. "Previously, this intelligence was found only in one or two data bases. Now, it will be accessed by almost everyone, such as the use of 800 numbers to make queries and conduct business transactions."

Robinson's "intelligent network services" will permit a business to operate from virtually any location in the world instead of restricting its operations from a single office or city. "In the future, small businesses will have the same communications features as larger businesses, and they are likely to be more competitive."

"For example," said Robinson, "a small businessman will be able to use a

The 1997 LCC and LCTX boards of directors were made up of (left to right, seated) Deborah D. Huffman, Douglas S. Shotwell, G.I. (Larry) Ross, Linwood D. Newman, Calvin W. Jayroe, and Douglass Shands, and (left to right, standing) Michael B. Robinson, Dale Green, Kirk Ross, Carlos R. Hamilton, Jr., Thomas O. Shelton, III, Deane C. Watson, Jr., Stephen P. Moore, Dale L. Trimble, Suzanne Simons Laidlaw, William P. Bowers, and Kenneth R. Whitton. This photo was made shortly before the company was acquired by Texas Utilities, Inc.

telephone network to route his calls from a contractor or supplier even while he is on vacation on the other side of the world. No longer does his business have to stop or slow down just because he isn't there," he said.

Robinson sees a growing use of E-mail, facsimile transmissions (FAX), and voice mail linked with networks. "Small businesses will no longer use answering machines, but will depend on networks and the Internet to do business from anywhere," he said.

This convergence of E-mail, FAXes, and voice mail will create what Robinson calls "asynchronous communications."

"Instead of spending a lot of time chatting about the weather, we're likely to be more direct and brief in our conversations because, most of the time, we won't be talking with people; we'll be leaving messages, and they'll respond with messages. The time spent in actually talking directly with people will decline."

Another major trend, Robinson says, will be data communications. "The Internet is making it available to anyone with a computer, and that means businesses are likely to have more interaction with their customers. People will tell businesses what they need, how to deliver products, and how to complete the payment. All of this will likely require less secretaries and clerks who have handled these chores in the past."

The growing popularity of the Internet has driven businesses to use more data transmissions to sell, deliver, and charge for products. In 1997, LCC moved to enhance its services to customers with Internet interests by acquiring Internet Publishing Technologies, formerly of Bryan/College Station. The company has grown since 1994 into a highly respected Internet business specializing in interactive web site development and hosting. Ross said the company will help broaden LCC's customer services, particularly to businesses.

The network Robinson envisions will, he feels, lead to more business activity from the home, breaking down traditional office boundaries.

The 1997 LCC and LCTX officers who served the companies before they were acquired by Texas Utilties, Inc., were (left to right), Michael B. Robinson, Dale Green, Kirk Ross, G.I. (Larry) Ross, Linwood D. Newman, Stephen P. Moore, Deborah D. Huffman, and Kenneth R. Whitton.

The 1990 Lufkin telephone directory opened the decade with a modern appearance.

While businesses will require more intelligence in their communications systems, Robinson feels the greatest advance among residential telephone customers will be the expansion of bandwidth in telephone lines. "The greater the bandwidth, the richer the applications are likely to be for our home customers," he said.

The big requirements for residential customers are likely to be voice service, Internet access and data delivery, and entertainment, such as video.

"The technology is already in place to allow people at home to call up news broadcasts on their computers, but there must be a greater acceptance before we see widespread usage of video via telephone lines," he said.

Digital subscriber lines, he said, will help facilitate faster access to data delivery in the home and to supply alternatives to cable television, but a networking service will have to be developed to expedite these services. "But, at some point in time, there will be three sources for entertainment in the home, cable TV, broadcast satellites, and telephone data deliveries. Accordingly, there will be fewer barriers to competition," he said

Robinson feels some of the advances in communications will force people to use their minds to a greater degree because they will be forced to think before they write. He also feels E-mail will broaden relationships because, as a service on the Internet, communications will be less costly and easier to send. "Written messages, in the years ahead, won't cost a dime," he said.

Spinoffs from home communications technology could include homebound schooling for children, graduate studies, church services delivered to the home at any time of the day or night, video E-mailing, and games played between Internet players miles and even nations apart.

The growing penetration of the communications market by wireless devices such as cellular telephones and pagers will skyrocket, Robinson predicts, and more E-mail will go wireless.

While technology exists for Robinson's predictions, he feels acceptance has yet to rise for many communications products. "Until the acceptance level rises to about 20 percent, some products are not likely to take off as quickly as others," he said. "Much of what we will be able to provide in the Lufkin-Conroe communications system will depend on this acceptance."

To many of the telephone company employees whose careers were intertwined with the growth of the Lufkin and Conroe exchanges, the new technology is difficult to fathom. "Things are moving so fast," said Sue Roberts of Lufkin, "that I'm having a hard time just keeping up with the telephone products we're offering today."

Looking back at the 100 years that have transpired since Dr. Denman and

Judge Mantooth strung a primitive telephone line between their offices, Lufkin-Conroe Telephone Exchange, Inc. and its affiliated companies clearly deserve their position as the fourth largest non-Bell telephone company in Texas and the 20th largest non-Bell system in the United States.

Much of the credit for the company's success lies in its embracing of new technology at critical junctures in 100 years, from a common battery switchboard in 1924 to fiber optics and networking innovations in the 1990s, and its ability to respond to the opportunities presented by a changing communications environment.

Ross summed up the company's progressive approach when he addressed an assembly of descendants of Dr. Denman and Judge Mantooth:

"Your grandparents gave us a beginning and a vision. We were fortunate to enhance that vision, only because we learned from them that to stand still in the face of a changing world would have been the wrong thing to do. If we had not embraced the vision of Dr. Denman and Judge Mantooth, and responded to the changing telephone environment, I doubt we would be celebrating a hundred years of continuous service."

G. I. (Larry) Ross, left, Chief Executive Officer of Lufkin-Conroe Communications Co., and Herb Zureich, President of TU Communications, conclude the merger of LCC with Texas Utilities.

22 The start of a new era

On the eve of the 100th anniversary of the inauguration of telephone service at Lufkin by Dr. Alexander Madison Denman and Judge Edwin James Mantooth, their descendants took a step that will significantly change the Lufkin-Conroe Communications Co. and its subsidiary companies in the coming century.

Shareholders of the company gave their approval to sell the company to Texas Utilities, Inc. on November 15, 1997. Company directors had previously approved the sale for $328 million on August 23, 1997.

The entire Lufkin-Conroe telephone operations will become a wholly-owned subsidiary of Texas Utilities and will retain its name in its traditional service territories.

On December 5, Erle Nye of Dallas, chairman of the board and chief executive offier of Texas Utilitiies, paid special tribute to the history and contributions of the Lufkin-Conroe system in a letter to Linwood D. Newman, chairman of the board of Lufkin-Conroe Communications Co. Mr. Nye's letter follows:

Dear Linwood:

It is not often that Texas Utilities Company has the opportunity to become associated with a company with the rich and remarkable history of Lufkin-Conroe Communications Co.

Clearly, the merger of our company and LCC does not mean the end of that history, but an extension of that history into an exciting new era for both LCC and Texas Utilities.

Drawing upon the experience and history of LCC and its people, we at TU envision breaking new ground in the field of communications and utility services. What LCC has achieved in East Texas since its founding 100 years ago will become the foundation for new and exciting things we expect to happen in the years ahead. Together, we can do something worthwhile for Texas.

In the coming months, we hope that each member of the LCC team will establish a personal relationship with our folks at TU and develop a sense of ownership and attachment.

Erle Nye, chairman of the board and chief executive officer of Texas Utilities Company.

We also encourage LCC's people to share with us the philosophy and skills that have made your company one of the country's most successful telephone companies. We hope to learn much from you.

We also want LCC to remain a good neighbor to the communities it serves and continue with the traditions of integrity, respect and customer service established decades ago by the Denman and Mantooth families.

Together, as we write another chapter in the histories of our two great companies, I hope all of us will continue to remind ourselves that our first priority is to enhance the lives of our customers with essential services they can enjoy well into the next century.

Thank you for the opportunity to let Texas Utilities Company help you write another chapter in LCC's wonderful history, and congratulations on your 100th anniversary.

Sincerely,

Erle Nye, Chairman

Herb Zureich, President of TU Communications, said of the acquisition by his company: "They (LCC) have been providing exemplary service in the Lufkin and Conroe areas for years by building an efficient, state of the art network and staffing it with energetic and highly-skilled employees, as well as actively applying new technologies and developing new services to anticipate competition and facilitate growth."

A merger of LCC and another company was inevitable, said LCC Chief Executive Officer G.I. (Larry) Ross.

"I've always felt if we were to be a viable player in the telecommunications industry's changing environment, we would have to be associated with a large, strong company capable of supporting the kind of investment that will be required in the industry during the years ahead," he said.

Ross said two laws—the Texas Public Utility Regulatory Act of 1995 and the Federal Telecommunications Act of 1996—changed the landscape of electronic communications and opened the door for increased competition in the industry. "With the passage of these two bills, it became evident that the convergence of the utility business—the merging of companies involved in electricity, natural gas, telecommunications, cable television, and other areas—would become more prevalent in the coming years," he said.

Another reason for the LCC-Texas Utilities merger, Ross said, was the growing need for investment capital by the Lufkin-Conroe company. "Our capital budgets were running from $21 to $25 million a year and for a company our size, that's a significant amount. To be a player in the coming telecommunications environment, it would have been necessary to borrow much more money, which may have necessitated the sale of stock and perhaps the dilution of the present shareholders' interest," he said.

"There are a significant number of business opportunities in our industry and, without adequate resources, we would have had to pass them by," said Ross.

The merger with Texas Utilities will not only provide the resources and support for expansion for LCC and its subsidiary companies, but will enable the company to compete with larger telecommunications companies such as Southwestern Bell, AT&T, and GTE.

In the coming years, Ross expects LCC and its employees will play a major role in the convergence of utility services as Texas Utilities broadens its services and consumer base. "As an example, in the future, residential and commercial developers are likely to look to a single source, like Texas Utilities, as a supplier for all utilities such as electricity, natural gas, communications, data transmission, video services, and even security," he explained.

The merger with Texas Utilities will result in few changes in the Lufkin and Conroe areas, Ross feels. The company will maintain its offices in the two cities because its communications systems and its technical personnel are embedded there. "We have a wealth of experienced, qualified people and I think we'll see additions to our staffs as we embark on new ventures in telecommunications," he said.

Rather than being the end of 100 years of growth in the telecommunications industry, Ross sees the merger with Texas Utilities as the start of another century. "And it is likely to be far more exciting and far-reaching than the last 100 years," he said.

While the Denman and Mantooth families will no longer be the principal owners of the company their grandfathers created, they will have the pride of knowing that their work of 100 years made a remarkable contribution to Lufkin, Conroe, and the telecommunications industry of Texas and the United States.

Research sources

Publications

Mrs. Ila Rice Edgar, University of Houston, masters thesis, 1941, document on file at Heritage Museum of Montgomery County, Conroe, Texas.

"History of Montgomery County," Montgomery County Genealogical Society, 1981, Hunter Publishing Company, Winston-Salem, North Carolina; Mary Peoples, book chairman.

"The New Handbook of Texas," Texas Historical Association, 1996, Austin, Texas, Ron G.Tyler, editor in chief.

"Hello, Texas. A History of Telephony in the Lone Star State," published by the Texas Telephone Association, Austin, Texas, 1990, edited by Jerry Hall.

"Land of The Little Angel, A History of Angelina County, " 1976, Lufkin Printing Company, edited by Bob Bowman.

"The Lufkin That Was," by Bob Bowman, 1982, Lufkin Printing Company.

"History of Angelina County," 1992, Angelina County Genealogical and Historical Society.

"The Physicians: 100 Years of Dedication," 1982, Angelina County Medical Society, compiled by Margie Kirkland.

Archival Materials

Archives, Lufkin-Conroe Telephone Exchange, Inc., Lufkin, Texas.

Various documents Kurth Memorial Library, Lufkin, Texas.

Various documents, Montgomery County Library, Conroe, Texas.

City Commission minutes, City of Lufkin, Lufkin, Texas.

City Council minutes, City of Conroe, Conroe, Texas.

Various documents, Heritage Museum of Montgomery County, Conroe, Texas.

Obituary records, Gipson Funeral Home, Lufkin, Texas.

Newspapers and Periodicals

Various articles, *The Conroe Courier*, Conroe, Texas,

Various articles, *The Lufkin Daily News*, Lufkin, Texas

Various articles, *The Houston Chronicle*, Lufkin, Texas.

Various articles, *The Houston Post*, Lufkin, Texas.

Various articles, *Telephone Engineering and Management.*

Various articles, *The Diboll Free Press*, Diboll, Texas.

Recollection notes, Jean Abbott, retired, Lufkin-Conroe Telephone Exchange, Inc.

Various articles, *Stromberg-Carlson Transmitter*, Stromberg-Carlson Company.

Various articles, *Look at Lufkin*, Lufkin, Texas.

Article on early telephone operators, Lufkin, Texas, Margie Kirkland, Lufkin, Texas.

Various articles, *The Connection*, Lufkin-Conroe Communications, Inc., Lufkin, Texas.

Interviews

Interview, Morris Bateman of Conroe, Texas, retired, Lufkin-Conroe Telephone Exchange, Inc., 1997, Conroe, Texas.

Interview, Larry Ross, chairman, Lufkin-Conroe Telephone Exchange, Inc. 1997, Lufkin, Texas.

Interview, Dr. Peyton Denman, shareholder, Lufkin-Conroe Communications, Inc., 1997, Lufkin, Texas.

Interview, Lucy Hendrick, retired telephone operator, Lufkin-Conroe Telephone Exchange, 1997, Lufkin, Texas.

Interview, Sue Roberts, employee, Lufkin-Conroe Telephone Exchange, Inc., 1997, Lufkin, Texas.

Interview, Walter Lawrence, retired employee, Lufkin-Conroe Telephone Exchange, Inc., 1997, Lufkin, Texas.

Interview, Betty Warren, retired employee, Lufkin-Conroe Telephone Exchange, Inc., 1997, Lufkin, Texas.

Interview, James T. (Jack) Mott, retired employee, Lufkin-Conroe Telephone Exchange, Inc., 1997, Lufkin, Texas.

Interview, William C. Tipson, retired employee, Lufkin-Conroe Telephone Exchange, Inc., 1997, Apple Springs, Texas.

Interview, Vera Cotton Acrey, retired employee, Lufkin-Conroe Telephone Exchange, Inc., 1997, Conroe, Texas.

Interview, Mike Robinson, Vice President, Marketing and Planning, Lufkin-Conroe Telephone Exchange, Inc., 1997, Lufkin, Texas.

Interview, Clifford and Peggy Lowry, retired employees, Lufkin-Conroe Telephone Exchange, Inc., 1997, Lufkin, Texas.

Interview, Dale Green, president, Lufkin-Conroe Telephone Exchange, Inc., 1997, Lufkin, Texas.

Interview, Doris Johns, retired employee, Lufkin-Conroe Telephone Exchange, Inc., 1997, Lufkin, Texas.

Interview, Pauline Bynum, retired employees, Lufkin-Conroe Telephone Exchange, Inc., 1997, Alto, Texas.

Interview, Mantooth (Bug) Shotwell, former employee, Lufkin-Conroe Telephone Exchange, Inc., 1997, Lufkin, Texas.

Interview, H.J. (Bubba) Shands, retired president, First Bank and Trust Company, 1997, Lufkin, Texas.

Photo Sources

Heritage Museum of Montgomery County, Conroe, Texas.

Museum of East Texas, Lufkin, Texas.

Kurth Memorial Library, Lufkin, Texas.

Montgomery County Library, Conroe, Texas.

Photo archives, Lufkin-Conroe Telephone Exchange, Inc., Lufkin, Texas.

Bob Bowman Photo Collection, Bob Bowman & Associates, Lufkin, Texas.

Index